Praise for *An Edible History of Humanity*

Selected by Indie Booksellers for the May 2009 Indie Next List

"Standage succeeds in underscoring the crucial role that food continues to play in our lives. Thousands of years ago, the invention of agriculture shaped early societies. Today, it connects us to global debates about trade and the environment. The book is further proof of the gastronomist Brillat-Savarin's truism, 'Tell me what you eat, and I will tell you what you are.'"

—Jane Black, *Washington Post*

"A fascinating history of the role of food in causing, enabling and influencing successive transformations of human society. This is an extraordinary and well-told story, a much neglected dimension to history."

—Sir Crispin Tickell, *Financial Times*

"The emphasis on food as a cultural catalyst differentiates Standage from Michael Pollan, whose plants' eye view of the world keeps the consumables central. With Standage it is not what changes in food that matters, but rather what food changes. And it's not just one food lifting and guiding history, but what Adam Smith might have called the 'invisible fork' of food economics."

—*New Scientist*

"Engaging and clear, this book will have a broad appeal to all who are interested in how food production, availability and trade have influenced the world, socially, economically and politically, and how food remains at the root of our future."

—*New Agriculturist*

"Farmers banded together and exploited their excess crops as a means of trade and currency. This allowed some people to abandon agriculture [leading to] organized communities and cities. Standage traces this ever-evolving story through Europe, Asia, and the Americas and casts human progress as an elaboration and refinement of this foundation . . . Standage also uncovers the aspects of food distribution that underlay such historic events as the Napoleonic Wars and the fall of the Soviet empire."

—*Booklist*

"[Standage] shows how one of humanity's most vital needs (hunger) didn't simply reflect but served as the driving force behind transformative and key events in history . . . Perhaps the most interesting section is the final one, which looks at the ways in which modern agricultural needs have acted as a spur for technological advancement, with Standage providing a summary of the challenges still faced by the green revolution."

—*Library Journal*

"This meaty little volume . . . 'concentrates specifically on the intersections between food history and world history.' But history isn't Standage's only concern. He takes the long view to illuminate and contextualize such contemporary issues as genetically modified foods, the complex relationship between food and poverty, the local food movement, the politicization of food and the environmental outcomes of modern methods of agriculture . . . Cogent, informative and insightful."

—*Kirkus Reviews*

An EDIBLE HISTORY *of* HUMANITY

TOM STANDAGE

BLOOMSBURY

NEW YORK · LONDON · OXFORD · NEW DELHI · SYDNEY

Bloomsbury USA
An imprint of Bloomsbury Publishing Plc

1385 Broadway 50 Bedford Square
New York London
NY 10018 WC1B 3DP
USA UK

www.bloomsbury.com

First published 2009
This paperback edition published 2010

© Tom Standage, 2009

ISBN: HB: 978-0-8027-1588-3
PB: 978-0-8027-1991-1
ePub: 978-0-8027-1982-9

Library of Congress Cataloging-in-Publication Data is available.

19

Book design by Simon M. Sullivan
Typeset by Westchester Book Group
Printed and bound in the U.S.A. by Sheridan, Chelsea, MI.

To Kirstin, my partner in food—and everything else

CONTENTS

INTRODUCTION
INGREDIENTS OF THE PAST

There is no history of mankind, there are only many histories of all kinds of aspects of human life.

—KARL POPPER

The fate of nations hangs upon their choice of food.

—JEAN-ANTHELME BRILLAT-SAVARIN

There are many ways to look at the past: as a list of important dates, a conveyor belt of kings and queens, a series of rising and falling empires, or a narrative of political, philosophical, or technological progress. This book looks at history in another way entirely: as a series of transformations caused, enabled, or influenced by food. Throughout history, food has done more than simply provide sustenance. It has acted as a catalyst of social transformation, societal organization, geopolitical competition, industrial development, military conflict, and economic expansion. From prehistory to the present, the stories of these transformations form a narrative that encompasses the whole of human history.

Food's first transformative role was as a foundation for entire civilizations. The adoption of agriculture made possible new settled lifestyles and set mankind on the path to the modern world. But the staple crops that supported the first civilizations—barley and wheat in the Near East, millet and rice in Asia, and maize and potatoes in the Americas—were not simply discovered by chance. Instead, they emerged through a complex process of coevolution, as desirable traits were selected and propagated by early farmers. These staple crops are, in effect, inventions: deliberately cultivated technologies that only exist as a result of human intervention. The story of the adoption of agriculture is the tale of how ancient genetic engineers developed powerful new tools that made civilization itself possible. In the process, mankind changed plants, and those plants in turn transformed mankind.

Having provided the platform on which civilizations could be founded, food subsequently acted as a tool of social organization, helping to shape and structure the complex societies that emerged. The political, economic, and religious structures of ancient societies, from hunter-gatherers to the first civilizations, were based upon the systems of food production and distribution. The production of agricultural food surpluses and the development of communal food-storage and irrigation systems fostered political centralization; agricultural fertility rituals developed into state religions; food became a medium of payment and taxation; feasts were used to garner influence and demonstrate status; food handouts were used to define and reinforce power structures. Throughout the ancient world, long before the invention of money, food was wealth—and control of food was power.

Once civilizations had emerged in various parts of the world, food helped to connect them together. Food-trade routes acted as international communications networks that fostered not just commercial exchange, but cultural and religious exchange too. The spice routes that spanned the Old World led to cross-cultural fertilization in fields as diverse as architecture, science, and religion. Early geographers started to take an interest in the customs and peoples of distant lands and compiled the first attempts at world maps. By far the greatest transformation caused by food trade was a result of the European desire to circumvent the Arab spice monopoly. This led to the discovery of the New World, the opening of maritime trade routes between Europe, America, and Asia, and the establishment by European nations of their first colonial outposts. Along the way, it also revealed the true layout of the world.

As European nations vied to build global empires, food helped to bring about the next big shift in human history: a surge in economic development through industrialization. Sugar and potatoes, as much as the steam engine, underpinned the Industrial Revolution. The production of sugar on plantations in the West Indies was arguably the earliest prototype of an industrial process, reliant though it was on slave labor. Potatoes, meanwhile, overcame initial suspicion among Europeans to become a staple food that produced more calories than cereal crops could from a given area of land. Together, sugar and potatoes provided cheap sustenance for the workers in the new factories of the industrial age. In Britain, where this process first began, the vexed question of whether the country's future lay in agriculture or in industry was unexpectedly and decisively resolved by the Irish Potato Famine of 1845.

The use of food as a weapon of war is timeless, but the large-scale military conflicts of the eighteenth and nineteenth centuries elevated it to a new level. Food played an important role in determining the outcome of the two wars that defined the United States of America: the Revolutionary War of the 1770s to 1780s and the Civil War of the 1860s. In Europe, meanwhile, Napoleon's rise and fall was intimately connected with his ability to feed his vast armies. The mechanization of warfare in the twentieth century meant that for the first time in history, feeding machines with fuel and ammunition became a more important consideration than feeding soldiers. But food then took on a new role, as an ideological weapon, during the Cold War between capitalism and communism, and ultimately helped to determine the outcome of the conflict. And in modern times food has become a battlefield for other issues, including trade, development, and globalization.

During the twentieth century the application of scientific and industrial methods to agriculture led to a dramatic expansion in the food supply and a corresponding surge in the world population. The so-called green revolution caused environmental and social problems, but without it there would probably have been widespread famine in much of the developing world during the 1970s. And by enabling the food supply to grow more rapidly than the population, the green revolution paved the way for the astonishingly rapid industrialization of Asia as the century drew to a close. Since people in industrial societies tend to have fewer children than those in agricultural societies, the peak in the human population, toward the end of the twenty-first century, is now in sight.

The stories of many individual foodstuffs, of food-related customs and traditions, and of the development of particular national cuisines have already been told. Less attention has been paid to the question of food's world-historical impact. This account does not claim that any single food holds the key to understanding history; nor does it attempt to summarize the entire history of food, or the entire history of the world. Instead, by drawing on a range of disciplines, including genetics, archaeology, anthropology, ethnobotany, and economics, it concentrates specifically on the intersections between food history and world history, to ask a simple question: which foods have done the most to shape the modern world, and how? Taking a long-term historical perspective also provides a new way to illuminate modern debates about food, such as the controversy surrounding genetically modified organisms, the relationship between food and poverty, the rise of the "local" food movement, the use of crops to make biofuels, the effectiveness of food as a means of mobilizing political support for various causes, and the best way to reduce the environmental impact of modern agriculture.

In his book *The Wealth of Nations,* first published in 1776, Adam Smith famously likened the unseen influence of market forces, acting on participants who are all looking out for their own best interests, to an invisible hand. Food's influence on history can similarly be likened to an invisible fork that has, at several crucial points in history, prodded humanity and altered its destiny, even though people were generally unaware of its influence at the time. Many food choices made in the past turn out to have had far-reaching consequences, and to have helped in unexpected ways to shape the world in which we now live. To the discerning eye, food's historical influence can be seen all around us, and not just in the kitchen, at the dining table, or in the supermarket. That food has been such an important ingredient in human affairs might seem strange, but it would be far more surprising if it had not: after all, everything that every person has ever done, throughout history, has literally been fueled by food.

PART I

THE EDIBLE FOUNDATIONS
OF CIVILIZATION

I

THE INVENTION OF FARMING

I have seen great surprise expressed in horticultural works at the wonderful skill of gardeners, in having produced such splendid results from such poor materials; but the art has been simple, and as far as the final result is concerned, has been followed almost unconsciously. It has consisted in always cultivating the best-known variety, sowing its seeds, and, when a slightly better variety chanced to appear, selecting it, and so onwards.

—CHARLES DARWIN, *The Origin of Species*

FOODS AS TECHNOLOGIES

What embodies the bounty of nature better than an ear of corn? With a twist of the wrist it is easily plucked from the stalk with no waste or fuss. It is packed with tasty, nutritious kernels that are larger and more numerous than those of other cereals. And it is surrounded by a leafy husk that shields it from pests and moisture. Maize appears to be a gift from nature; it even comes wrapped up. But appearances can be deceptive. A cultivated field of maize, or any other crop, is as man-made as a microchip, a magazine, or a missile. Much as we like to think of farming as natural, ten thousand years ago it was a new and alien development. Stone Age hunter-gatherers would have regarded neatly cultivated fields, stretching to the horizon, as a bizarre and unfamiliar sight. Farmed land is as much a technological landscape as a biological one. And in the grand scheme of human existence, the technologies in question—domesticated crops—are very recent inventions.

The ancestors of modern humans diverged from apes about four and a half million years ago, and "anatomically modern" humans emerged around 150,000 years ago. All of these early humans were hunter-gatherers who subsisted on plants and animals that were gathered and hunted in the wild. It is only within the past 11,000 years or so that humans began to cultivate food deliberately. Farming emerged independently in several different times and places, and had taken hold in the Near East by around 8500 B.C., in China by around 7500 B.C., and in Central and South America by around 3500 B.C. From these three main starting points, the technology of farming then spread throughout the world to become mankind's chief means of food production.

This was a remarkable change for a species that had relied on a nomadic lifestyle based on hunting and gathering for its entire previous existence. If the 150,000 years since modern humans emerged are likened to one hour, it is only in the last four and a half minutes that humans began to adopt farming, and agriculture only became the dominant means of providing human subsistence in the last minute and a half. Humanity's switch from foraging to farming, from a natural to a technological means of food production, was recent and sudden.

Though many animals gather and store seeds and other foodstuffs, humans are unique in deliberately cultivating specific crops and selecting and propagating particular desired characteristics. Like a weaver, a carpenter, or a blacksmith, a farmer creates useful things that do not occur in nature. This is done using plants and animals that have been modified, or domesticated, so that they better suit human purposes. They are human creations, carefully crafted tools that are used to produce food in novel forms, and in far greater quantities than would occur naturally. The significance of their development cannot be overstated, for they literally made possible the modern world. Three domesticated plants in particular—wheat, rice, and maize—proved to be most significant. They laid the foundations for civilization and continue to underpin human society to this day.

THE MAN-MADE NATURE OF MAIZE

Maize, more commonly known as corn in America, provides the best illustration that domesticated crops are unquestionably human creations. The distinction between wild and domesticated plants is not a hard and fast one. Instead, plants occupy a continuum: from entirely wild plants, to domesticated ones that have had some characteristics modified to suit humans, to entirely domesticated plants, which can only reproduce with human assistance. Maize falls into the last of these categories. It is the result of human propagation of a series of random genetic mutations that transformed it from a simple grass into a bizarre, gigantic mutant that can no longer survive in the wild. Maize is descended from teosinte, a wild grass indigenous to modern-day Mexico. The two plants look very different. But just a few genetic mutations, it turns out, were sufficient to transform one into the other.

One obvious difference between teosinte and maize is that teosinte ears consist of two rows of kernels surrounded by tough casings, or glumes, which protect the edible kernels within. A single gene, called *tga1* by modern geneticists, controls the size of these glumes, and a mutation in the gene results in exposed kernels. This means the kernels are less likely to survive the journey through the digestive tract of an animal, placing mutant plants at a reproductive disadvantage to nonmutants, at least in the normal scheme of things. But the exposed kernels would also have made teosinte far more attractive to human foragers, since there would have been no need to remove the glumes before consumption. By gathering just the mutant plants with exposed kernels, and then sowing some of them as seeds, proto-farmers could increase the proportion of plants with exposed kernels. The *tga1* mutation, in short, made teosinte plants less likely to survive in the wild, but also made them more attractive to humans, who propagated the mutation. (The glumes in maize are so reduced that you only notice them today when they get stuck between your teeth. They are the silky, transparent film that surrounds each kernel.)

*Progression from teosinte to proto-maize
and modern maize.*

Another obvious difference between teosinte and maize lies in the overall structure, or architecture, of the two plants, which determines the position and number of the male and female reproductive parts, or inflorescences. Teosinte has a highly branched architecture with multiple stalks, each of which has one male inflorescence (the tassel) and several female inflorescences (the ears). Maize, however, has a single stalk with no branches, a single tassel at the top, and far fewer but much larger ears halfway up the stalk, enclosed in a leafy husk. Usually there is just one ear, but in some varieties of maize there can be two or three. This change in architecture seems to be the result of a mutation in a gene known as *tb1*. From the plant's point of view, this mutation is a bad thing: It makes fertilization, in which pollen from the tassel must make its way down to the ear, more difficult.

But from the point of view of humans, it is a very helpful mutation, since a small number of large ears is easier to collect than a large number of small ones. Accordingly, proto-farmers would have been more likely to gather ears from plants with this mutation. By sowing their kernels as seeds, humans propagated another mutation that resulted in an inferior plant, but a superior food.

The ears, being closer to the ground, end up closer to the nutrient supply and can potentially grow much larger. Once again, human selection guided this process. As proto-farmers gathered ears of proto-maize, they would have given preference to plants with larger ears; and kernels from those ears would then have been used as seeds. In this way, mutations that resulted in larger ears with more kernels were propagated, so that the ears grew larger from one generation to the next and became corn cobs. This can clearly be seen in the archaeological record: At one cave in Mexico, a sequence of cobs has been found, increasing in length from a half inch to eight inches long. Again, the very trait that made maize attractive to humans made it less viable in the wild. A plant with a large ear cannot propagate itself from one year to the next, because when the ear falls to the ground and the kernels sprout, the close proximity of so many kernels competing for the nutrients in the soil prevents any of them from growing. For the plant to grow, the kernels must be manually separated from the cob and planted a sufficient distance apart— something only humans can do. As maize ears grew larger, in short, the plant ended up being entirely dependent on humans for its continued existence.

What started off as an unwitting process of selection eventually became deliberate, as early farmers began to propagate desirable traits on purpose. By transferring pollen from the tassel of one plant to the silks of another, it was possible to create new varieties that combined the attributes of their parents. These new varieties had to be kept away from other varieties to prevent the loss of desirable traits. Genetic analysis suggests that one particular type of teosinte, called Balsas teosinte, is most likely to have been the progenitor of maize. Further analysis of regional varieties of Balsas teosinte suggests that maize was originally domesticated in central Mexico, where the modern-day states of Guerrero, México, and Michoacán meet. From here, maize spread and became a staple food for peoples throughout the Americas: the Aztecs and Maya of Mexico, the Incas of Peru, and many other tribes and cultures throughout North, South, and Central America.

But maize could only become a dietary mainstay with the help of a further technological twist, since it is deficient in the amino acids lysine and tryptophan, and the vitamin niacin, which are essential elements of a healthy human diet. When maize was merely one foodstuff among many these deficiencies did not matter, since other foods, such as beans and squash, made up for them. But a maize-heavy diet results in pellagra, a nutritional disease characterized by nausea, rough skin, sensitivity to light, and dementia. (Light sensitivity due to pellagra is thought to account for the origin of European vampire myths, following the introduction of maize into European diets in the eighteenth century.) Fortunately, maize can be rendered safe by treating it with calcium hydroxide, in the form of ash from burnt wood or crushed shells, which is either added directly to the cooking pot, or mixed with water to create an alkaline solution in which the maize is left to soak overnight. This has the effect of softening the kernels and making them easier to prepare, which probably explains the origin of the practice. More importantly but less visibly, it also liberates amino acids and niacin, which exist in maize in an inaccessible or "bound" form called niacytin. The resulting processed kernels were called *nixtamal* by the Aztecs, so that the process is known today as nixtamalization. This practice seems to have been developed as early as 1500 B.C.; without it, the great maize-based cultures of the Americas could never have been established.

All of this demonstrates that maize is not a naturally occurring food at all. Its development has been described by one modern scientist as the most impressive feat of domestication and genetic modification ever undertaken. It is a complex technology, developed by humans over successive generations to the point where maize was ultimately incapable of surviving on its own in the wild, but could deliver enough food to sustain entire civilizations.

CEREAL INNOVATION

Maize is merely one of the most extreme examples. The world's two other major staples, which went on to underpin civilization in the Near East and Asia respectively, are wheat and rice. They too are the results of human selective processes that propagated desirable mutations to create more convenient and abundant foodstuffs. Like maize, both wheat and rice are cereal grains, and the key difference between their wild and domesticated forms is that domesticated varieties are "shatterproof." The grains are attached to a central axis known as the rachis. As the wild grains ripen the rachis becomes brittle, so that when touched or blown by the wind it shatters, scattering the grains as seeds. This makes sense from the plant's perspective, since it ensures that the grains are only dispersed once they have ripened. But it is very inconvenient from the point of view of humans who wish to gather them.

In a small proportion of plants, however, a single genetic mutation means the rachis does not become brittle, even when the seeds ripen. This is called a "tough rachis." This mutation is undesirable for the plants in question, since they are unable to disperse their seeds. But it is very helpful for humans gathering wild grains, who are likely to gather a disproportionate number of tough-rachis mutants as a result. If some of the grains are then planted to produce a crop the following year, the tough-rachis mutation will be propagated, and every year the proportion of tough-rachis mutants will increase. Archaeologists have demonstrated in field experiments with wheat that this is exactly what happens. They estimate that plants with tough, shatterproof rachises would become predominant within about two hundred years—which is roughly how long the domestication of wheat seems to have taken, according to the archaeological record. (In maize, the cob is in fact a gigantic shatterproof rachis.)

As with maize, proto-farmers selected for other desirable characteristics in wheat, rice, and other cereals during the process of domestication. A mutation in wheat causes the hard glumes that cover each grain to separate more easily, resulting in "self-threshing" varieties. The individual grains are less well protected as a result, so this mutation is bad news in the wild. But it is helpful to human farmers, since it makes it easier to separate the edible grains after beating sheaves of cut wheat on a stone threshing floor. When grains were being plucked from the floor, small grains and those with glumes still attached would have been passed over in favor of larger ones without glumes. This helped to propagate these helpful mutations.

Another trait common to many domesticated crops is the loss of seed dormancy, the natural timing mechanism that determines when a seed germinates. Many seeds require specific stimuli, such as cold or light, before they will start growing, to ensure that they only germinate under favorable circumstances. Seeds that remain dormant until after a cold spell, for example, will not germinate in the autumn, but will wait until after the winter has passed. Human farmers would often like seeds to start growing as soon as they are planted, however. Given a collection of seeds, some of which exhibit seed dormancy and some of which do not, it is clear that those that start growing right away stand a better chance of being gathered and thus forming the basis of the next crop. So any mutations that suppress seed dormancy will tend to be propagated.

Similarly, wild cereals germinate and ripen at different times. This ensures that whatever the pattern of rainfall, at least some of the grains will mature to provide seeds for the following year. Harvesting an entire field of grain on the same day, however, favors grains that are almost ripe at the time. Grains that are over-ripe or under-ripe will be less viable if sown as seeds the following year. The effect is to reduce the variation in ripening time from one year to the next, so that eventually the entire field ripens at the same time. This is bad from the plant's point of view, since it means the entire crop can potentially fail. But it is far more convenient for human farmers.

In the case of rice, human intervention helped to propagate desirable properties such as taller and larger plants to aid harvesting, and more secondary branches and larger grains to increase yield. But domestication also made wheat and rice more dependent on human intervention. Rice lost its natural ability to survive in flood waters, for example, as it was pampered by human farmers. And both wheat and rice were less able to reproduce by themselves because of the human-selected shatterproof rachis. The domestication of wheat, rice, and maize, the three main cereal grains, and of their lesser siblings barley, rye, oats, and millet, were all variations on the same familiar genetic theme: more convenient food, less resilient plant.

The same trade-off occurred as humans domesticated animals for the purpose of providing food, starting with sheep and goats in the Near East around 8000 B.C. and followed by cattle and pigs soon afterward. (Pigs were independently domesticated in China at roughly the same time, and the chicken was domesticated in southeast Asia around 6000 B.C.) Most domesticated animals have smaller brains and less acute eyesight and hearing than their wild ancestors. This reduces their ability to survive in the wild but makes them more docile, which suits human farmers.

Humans became dependent on their new creations, and vice versa. By providing a more dependable and plentiful food supply, farming provided the basis for new lifestyles and far more complex societies. These cultures relied on a range of foods, but the most important were the cereals: wheat and barley in the Near East, rice and millet in Asia, and maize in the Americas. The civilizations that subsequently arose on these edible foundations, including our own, owe their existence to these ancient products of genetic engineering.

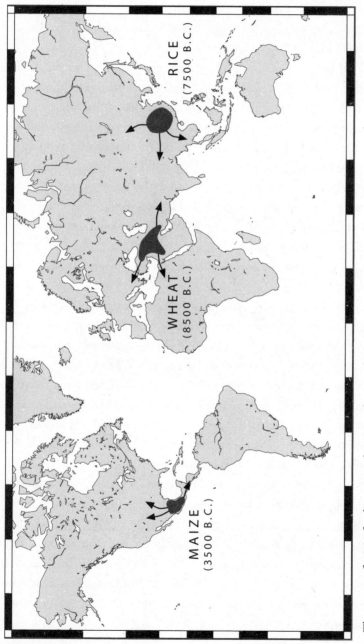

RICE
(7500 B.C.)

WHEAT
(8500 B.C.)

MAIZE
(3500 B.C.)

The centers of origin for domesticated maize, wheat, and rice.

Present at the Creation

This debt is acknowledged in many myths and legends in which the creation of the world, and the emergence of civilization after a long period of barbarism, are closely bound up with these vital crops. The Aztecs of Mexico, for example, believed men were created five times, each generation being an improvement over the last. Teosinte was said to have been man's principal food in the third and fourth creations. Finally, in the fifth creation, man nourished himself with maize. Only then did he prosper, and his descendants populated the world.

The creation story of the Maya of southern Mexico, recounted in the Popul Vuh (or "sacred book"), also involves repeated attempts to create mankind. At first the gods fashioned men out of mud, but the resulting creatures could barely see, could not move at all, and were soon washed away. So the gods tried again, this time making men out of wood. These creatures could walk on all fours and speak, but they lacked blood and souls, and they failed to honor the gods. The gods destroyed these men, too, so that all that remained of them were a few tree-dwelling monkeys. Finally, after much discussion about the appropriate choice of ingredients, the gods made a third generation of men from white and yellow ears of maize: "Of yellow maize and of white maize they made their flesh; of corn-meal dough they made the arms and the legs of man. Only dough of corn-meal went into the flesh of our first fathers, the four men, who were created." The Maya believed they were descended from these four men and their wives, who were created shortly afterward.

Maize also features in the story told by the Incas of South America to explain their origins. In ancient times, it is said, the people around Lake Titicaca lived like wild animals. The sun god, Inti, took pity on them and sent his son Manco Capac and his daughter Mama Ocllo, who were also husband and wife, to civilize them. Inti gave Manco Capac a golden stick with which to test the fertility of the soil and its suitability for growing maize. Having found a suitable place, they were to found a state and instruct its people in the proper worship of the sun god. The couple's travels finally brought them to the Cuzco Valley, where the golden stick disappeared into the ground. Manco Capac taught the people about farming and irrigation, Mama Ocllo taught them about spinning and weaving, and the valley became the center of the Inca civilization. Maize was regarded as a sacred crop by the Incas, even though potatoes also formed a large part of their diet.

Rice too appears in countless myths in the countries where it is grown. In Chinese myths, rice appears to save mankind when it is on the verge of starvation. According to one story, the goddess Guan Yin took pity on the starving humans and squeezed her breasts to produce milk, which flowed into the previously empty ears of the rice plants to become rice grains. She then pressed harder, causing a mixture of blood and milk to flow into some of the plants. This is said to explain why rice exists in both red and white varieties. Another Chinese tale tells of a great flood, after which very few animals remained for hunting. As they searched for food, the people saw a dog coming toward them with bunches of long, yellow seeds hanging from its tail. They planted the seeds, which grew into rice and dispelled their hunger forever. In a different series of rice myths, told in Indonesia and throughout the islands of Indochina, rice appears as a delicate and virtuous maiden. The Indonesian rice goddess, Sri, is the goddess of the earth who protects the people against hunger. One story tells how Sri was killed by the other gods to protect her from the lecherous advances of the king of the gods, Batara Guru. When her body was buried, rice sprouted from her eyes and sticky rice grew from her chest. Filled with remorse, Batara Guru gave these crops to mankind to cultivate.

The tale of the creation of the world and the emergence of civilization told by the Sumerians, the ancient inhabitants of what is now southern Iraq, refers to a time after the creation of the world by Anu, when people existed but agriculture was unknown. Ashnan, the grain goddess, and Lahar, the goddess of sheep, had not yet appeared; Tagtug, patron of the craftsmen, had not been born; and

Mirsu, the god of irrigation, and Sumugan, the god of cattle, had not arrived to help mankind. As a result, "the grain . . . and barley-grain for the cherished multitudes were not yet known." Instead, the people ate grass and drank water. The goddesses of grain and flocks were then created to provide food for the gods, but no matter how much the gods ate, they were not filled. Only with the emergence of civilized men, who made regular offerings of food to the gods, were the gods' appetites finally satisfied. So domesticated crops and animals were a gift to man that conferred upon him an obligation to make regular food offerings to the gods. This tale preserves a folk memory of a time before the adoption of farming, when humans were still foragers. Similarly, a Sumerian hymn to the grain goddess describes a barbaric age before cities, fields, sheepfolds, and cattle stalls—an era that came to an end when the grain goddess inaugurated a new era of civilization.

Contemporary explanations of the genetic basis of plant and animal domestication are really just the modern, scientific version of these ancient and strikingly similar creation myths from around the world. Today, we would say that the abandonment of hunting and gathering, the domestication of plants and animals, and the adoption of a settled lifestyle based on farming put mankind on the road to the modern world, and that those earliest farmers were the first modern, "civilized" humans. Admittedly, this is a rather less colorful account than those provided by the various creation myths. But given that the domestication of certain key cereal crops was an essential step toward the emergence of civilization, there is no doubt that these ancient tales contain far more than just a grain of truth.

2

THE ROOTS OF MODERNITY

Cursed is the ground because of you; through painful toil you will eat of it all the days of your life.

—GENESIS 3:17

AN AGRICULTURAL MYSTERY

The mechanism by which plants and animals were domesticated may be understood, but that does little to explain the motivations of the people in question. Quite why humans switched from hunting and gathering to farming is one of the oldest, most complex, and most important questions in human history. It is mysterious because the switch made people significantly worse off, from a nutritional perspective and in many other ways. Indeed, one anthropologist has described the adoption of farming as "the worst mistake in the history of the human race."

Compared with farming, being a hunter-gatherer was much more fun. Modern anthropologists who have spent time with surviving hunter-gatherer groups report that even in the marginal areas where they are now forced to live, gathering food only accounts for a small proportion of their time—far less than would be required to produce the same quantity of food via farming. The !Kung Bushmen of the Kalahari, for example, typically spend twelve to nineteen hours a week collecting food, and the Hazda nomads of Tanzania spend less than fourteen hours. That leaves a lot of time free for leisure activities, socializing, and so on. When asked by an anthropologist why his people had not adopted farming, one Bushman replied, "Why should we plant, when there are so many mongongo nuts in the world?" (Mongongo fruits and nuts, which comprise around half the !Kung diet, are gathered from wild stands of trees and are abundant even when no effort is made to propagate them.) In effect, hunter-gatherers work two days a week and have five-day weekends.

The hunter-gatherer lifestyle in preagricultural times, in less marginal environments, would probably have been even more pleasant. It used to be thought that the switch to farming gave people more time to devote to artistic pursuits, the development of new crafts and technologies, and so on. Farming, in this view, was a liberation from the anxious hand-to-mouth existence of the hunter-gatherer. But in fact the opposite turns out to be true. Farming is more productive in the sense that it produces more food per unit of land: a group of twenty-five people can subsist by farming on a mere twenty-five acres, a much smaller area than the tens of thousands of acres they would need to subsist by hunting and gathering. But farming is less productive when measured by the amount of food produced per hour of labor. It is, in other words, much harder work.

Surely this effort was worthwhile if it meant that people no longer needed to worry about malnutrition or starvation? So you might think. Yet hunter-gatherers actually seem to have been much healthier than the earliest farmers. According to the archaeological evidence, farmers were more likely than hunter-gatherers to suffer from dental-enamel hypoplasia—a characteristic horizontal striping of the teeth that indicates nutritional stress. Farming results in a less varied and less balanced diet than hunting and gathering does. Bushmen eat around seventy-five different types of wild plants, rather than relying on a few staple crops. Cereal grains provide reliable calories, but they do not contain the full range of essential nutrients.

So farmers were shorter than hunter-gatherers. This can be determined from skeletal remains by comparing the "dental" age derived from the teeth with the "skeletal" age implied by the lengths of the long bones. A skeletal age that is lower than the dental age is evidence of stunted growth due to malnutrition. Skeletal evidence from Greece and Turkey suggests that at the end of the last ice age, around 14,000 years ago, the average height of hunter-gatherers was five feet nine inches for men and five feet five inches for women. By 3000 B.C., after the adoption of farming, these averages had fallen to five feet three inches for men and five feet for women. It is only in modern times that humans have regained the stature of ancient hunter-gatherers, and only in the richest parts of the world. Modern Greeks and Turks are still shorter than their Stone Age ancestors.

In addition, many diseases damage bones in characteristic ways, and evidence from studies of bones reveals that farmers suffered from various diseases of malnutrition that were rare or absent in hunter-gatherers. These include rickets (vitamin D deficiency), scurvy (vitamin C deficiency), and anemia (iron deficiency). Farmers were also more susceptible to infectious diseases such as leprosy, tuberculosis, and malaria as a result of their settled lifestyles. And their dependence on cereal grains had other specific consequences: female skeletons often display evidence of arthritic joints and deformities of the toes, knees, and lower back, all of which are associated with the daily use of a saddle quern to grind grain. Dental remains show that farmers suffered from tooth decay, unheard of in hunter-gatherers, because the carbohydrates in the farmers' cereal-heavy diets were reduced to sugars by enzymes in their saliva as they chewed. Life expectancy, which can also be determined from skeletons, also fell: Evidence from the Illinois River Valley shows that average life expectancy at birth fell from twenty-six for hunter-gatherers to nineteen for farmers.

At some archaeological sites it is possible to follow health trends as hunter-gatherers become more sedentary and eventually adopt farming. As the farming groups settle down and grow larger, the incidence of malnutrition, parasitic diseases, and infectious diseases increases. At other sites, it is possible to compare the condition of hunter-gatherers and farmers living alongside each other. The settled farmers are invariably less healthy than their free-roaming neighbors. Farmers had to work much longer and harder to produce a less varied and less nutritious diet, and they were far more prone to disease. Given all these drawbacks, why on earth did people take up farming?

THE ORIGINS OF FARMING

The short answer is that they did not realize what was happening until it was too late. The switch from hunting and gathering to farming was a gradual one from the perspective of individual farmers, despite being very rapid within the grand scheme of human history. For just as wild crops and domesticated crops occupy a continuum, there is a range from pure hunter-gatherer to relying entirely on farmed foods.

Hunter-gatherers sometimes manipulate ecosystems to increase the availability of food, though such behavior falls far short of the deliberate large-scale cultivation we call farming. Using fire to clear land and prompt new growth, for example, is a practice that goes back at least 35,000 years. Australian aborigines, one of the few remaining groups of hunter-gatherers to have survived into modern times, plant seeds on occasion to increase the availability of food when they return to a particular site a few months later. It would be an exaggeration to call this farming, since such food makes up only a tiny fraction of their diet. But the deliberate manipulation of the ecosystem means they are not exclusively hunter-gatherers either.

The adoption of farming seems to have happened as people moved gradually along the spectrum from being pure hunter-gatherers to being ever more reliant on (and eventually dependent on) farmed food. Theories to explain this shift abound, but there was probably no single cause. Instead a combination of factors were probably involved, each of which played a greater or lesser role in each of the homelands where agriculture arose independently.

One of the most important factors appears to have been climate change. Studies of the ancient climate, based on the analysis of ice cores, deep-sea cores, and pollen profiles, have found that between

18,000 B.C. and 9500 B.C. the climate was cold, dry, and highly vari-able, so any attempt to cultivate or domesticate plants would have failed. Intriguingly there is evidence of at least one such attempt, at a site called Abu Hureyra in northern Syria. Around 10,700 B.C. the inhabitants of this site seem to have begun to domesticate rye. But their attempt fell victim to a sudden cold phase known as the Younger Dryas, which began around 10,700 B.C. and lasted for around 1,200 years. Then, around 9500 B.C., the climate suddenly became warmer, wetter, and more stable. This provided a necessary but not sufficient condition for agriculture. After all, if the newly stable climate was the only factor that prompted the adoption of farming, then people would have adopted it simultaneously all around the world. But they did not, so there must have been other forces at work as well.

One such factor was greater sedentism, as hunter-gatherers in some parts of the world became less mobile and began to spend most of the year at a single camp, or even took up permanent residence. There are many examples of sedentary village communities that predate the adoption of farming, such as those of the Natufian culture of the Near East, which flourished in the millennium before the Younger Dryas, and others on the north coast of Peru and in North America's Pacific Northwest. In each case these settlements were made possible by abundant local wild food, often in the form of fish or shellfish. Normally, hunter-gatherers move their camps to prevent the food sup-ply in a particular area from becoming depleted, or to take advantage of the seasonal availability of different foods. But there is no need to move around if you settle next to a river and the food comes to you. Improvements in food-gathering techniques in the late Stone Age, such as better arrows, nets, and fish hooks, may also have promoted sedentism. Once a hunter-gatherer band could extract more food (such as fish, small rodents, or shellfish) from its surroundings, it did not need to move around so much.

Sedentism does not always lead to farming, and some settled hunter-gatherer groups survived into modern times without ever adopting agriculture. But sedentism does make the switch to farming more likely. Settled hunter-gatherers who gather wild grains, for example, might be inclined to start planting a few seeds in order to maintain the supply. Planting might also have provided a form of insurance against variations in the supply of other foods. And since grains are processed using grinding stones which are inconvenient for hunter-gatherers to carry from one camp to another, greater sedentism would have made grains a more attractive foodstuff. The fact that grains are energy-rich, and could be dried and stored for long periods, also counted in their favor. They were not a terribly exciting foodstuff, but they could be relied upon in extremis.

It is not hard to imagine how sedentary hunter-gatherers might have started to rely more heavily on cereal grains as part of their diet. What was initially a relatively unimportant food gradually became more important, for the simple reason that proto-farmers could ensure its availability (by planting and subsequent storage) in ways they could not for other foods. Archaeological evidence from the Near East suggests that proto-farmers initially cultivated whatever wild cereals were at hand, such as einkorn wheat. But as they became more reliant on cereals they switched to more productive crops, such as emmer wheat, which produce more food for a given amount of labor.

Population growth as a result of sedentism has also been suggested as a contributory factor in the adoption of farming. Nomadic hunter-gatherers have to carry everything with them when they move camp, including infants. Only when a child can walk unaided over long distances, at the age of three of four, can its mother contemplate having another baby. Women in settled communities, however, do not face this problem and can therefore have more children. This would have placed greater demands on the local food supply and might have encouraged supplemental planting and, eventually, agriculture. One drawback with this line of argument, however, is that in some parts of the world the population density appears to have increased significantly only after the adoption of farming, not beforehand.

There are many other theories. In some parts of the world hunter-gatherers may have turned to farming as the big-game species that were their preferred prey declined in number. Farming may have been prompted by social competition, as rival groups competed to host the most lavish feasts; this might explain why, in some parts of the world, luxury foods appear to have been domesticated before staples. Or perhaps the inspiration was religious, and people planted seeds as a fertility rite, or to appease the gods after harvesting wild grains. It has even been suggested that the accidental fermentation of cereal grains, and the resulting discovery of beer, provided the incentive for the adoption of farming, in order to guarantee a regular supply.

The important thing is that at no point did anyone make a conscious decision to adopt an entirely new lifestyle. At every step along the way, people simply did what made the most sense at the time: Why be a nomad when you can settle down near a good supply of fish? If wild food sources cannot be relied upon, why not plant a few seeds to increase the supply? The proto-farmers' slowly increasing dependence on cultivated food took the form of a gradual shift, not a sudden change. But at some point an imperceptible line was crossed, and people began to become dependent on farming. The line is crossed when the wild food resources in the surrounding area, were they to be fully exploited, are no longer enough to sustain the population. The deliberate production of supplementary food through farming is then no longer optional, but has become compulsory. At this point there is no going back to a nomadic, hunter-gatherer lifestyle—or not, at least, without significant loss of life.

DID FARMERS SPREAD, OR DID FARMING SPREAD?

Farming then poses a second puzzle. Once agriculture had taken root in a few parts of the world, the question then becomes: Why did it spread almost everywhere else? One possibility is that farmers spread out, displacing or exterminating hunter-gatherers as they went. Alternatively, hunter-gatherers on the fringes of farming areas might have decided to follow suit and become farmers themselves, adopting the methods and the domesticated crops and animals of their farming neighbors. These two possibilities are known as "demic diffusion" and "cultural diffusion" respectively. So was it the actual farmers or merely the idea of farming that spread?

The idea that farmers spread out from the agricultural homelands, taking domesticated crops and knowledge of farming techniques with them as they went, is supported by evidence from many parts of the world. As farmers set out to establish new communities on unfarmed land, the result was a "wave of advance" centered on the areas where domestication first occurred. Greece appears to have been colonized by farmers who arrived by sea from the Near East between 7000 B.C. and 6500 B.C., for example. Archaeologists have found very few hunter-gatherer sites, but hundreds of early farming sites, in the country. Similarly, farmers arriving via the Korean peninsula from China seem to have introduced rice agriculture to Japan starting in around 300 B.C. Linguistic evidence also supports the idea of a migration from agricultural homelands in which languages, as well as farming practices, were dispersed. The distribution of language families in Europe, East Asia, and Austronesia is broadly consistent with the archaeological evidence for the diffusion of agriculture. Today, nearly 90 percent of the world's population speaks a language belonging to one of seven language families that had their origins in two agricultural homelands: the Fertile Crescent and parts of China. The languages we speak today, like the foods we eat, are descended from those used by the first farmers.

Yet there is also evidence to suggest that hunter-gatherers were not always pushed aside or exterminated by incoming farmers, but lived alongside them and in some cases became farmers too. The clearest example is found in southern Africa, where Khoisan hunter-gatherers adopted Eurasian cattle from the north and became herders. Several European sites provide archaeological evidence of farmers and hunter-gatherers living side by side and trading goods.

The two types of community had very different ideas about what sort of sites were desirable for settlement, so there is no reason why they could not have coexisted, as long as suitable ecological niches remained for hunter-gatherers. Things would have become progressively more difficult for hunter-gatherers living near farmers, however. Farmers would not have worried so much about overexploiting wild food resources near their settlements, given that they had farmed foods to fall back on. Eventually the hunter-gatherers either joined farming communities, or adopted farming themselves, or were forced to move to new areas.

So which mechanism predominated? In Europe, where the advent of farming has been most intensely studied, researchers have used genetic analysis to determine whether modern Europeans' ancestors were predominantly indigenous hunter-gatherers who took up farming or immigrant farmers who arrived from the Near East. In such studies, people from the Anatolian peninsula (western Turkey), which lies within the Fertile Crescent, are taken to be genetically representative of the earliest farmers. Similarly, Basques are assumed to be the most direct descendants of hunter-gatherers, for two reasons. First, the Basque language bears no resemblance to European languages descended from proto–Indo-European, the language family imported into Europe along with farming, and instead appears to date back to the Stone Age. (Several Basque words for tools begin with "aitz," the word for stone, which suggests that the words date from a time when stone tools were in use.) Second, there are several Basque-specific genetic variations that are not found in other Europeans.

In one recent study, genetic samples were taken from both these groups and were then compared with samples from populations in different parts of Europe. The researchers found that the genetic contributions from Basques and Anatolians varied significantly across Europe: The Anatolian (that is, Near Eastern farmer) contribution was 79 percent in the Balkans, 45 percent in northern Italy, 63 percent in southern Italy, 35 percent in southern Spain, and 21 percent in England. In short, the contribution from farmers was highest in the east and lowest in the west. And this provides the answer to the puzzle. It suggests that farming spread as a result of a hybrid process in which a migrant farming population spread into Europe from the east and was gradually diluted by intermarriage, so that the resulting population ended up being descended from both groups. The same thing probably happened in other parts of the world, too.

The spread of farming from its agricultural homelands, followed by the population growth of farming communities, meant that farmers outnumbered hunter-gatherers within a few thousand years. By 2000 B.C., the majority of humanity had taken up farming. This was such a fundamental change that even today, many thousands of years later, the distribution of human languages and genes continues to reflect the advent of farming. During domestication, plants were genetically reconfigured by humans; and as agriculture was adopted, humans were genetically reconfigured by plants.

Man, an Agricultural Animal

Human farmers and their domesticated plants and animals struck a grand bargain, though the farmers did not realize it at the time, and their fates became intertwined. Consider maize. Domestication made it dependent on man, but its alliance with humans also carried maize far beyond its origins as an obscure Mexican grass, so that it is now one of the most widely planted crops on earth. From mankind's point of view, meanwhile, the domestication of maize made available an abundant new source of food; but its cultivation (like that of other plants) prompted people to adopt a new, sedentary lifestyle based on farming. Is man exploiting maize for his own purposes, or is maize exploiting man? Domestication, it seems, is a two-way street.

Even today, thousands of years after the first farmers began the process of domesticating plants and animals, mankind is still a farming species, and food production remains humanity's primary occupation. Agriculture employs 41 percent of the human race, more than any other activity, and accounts for 40 percent of the world's land area. (About a third of this land is used for crop production, and about two thirds provide pasture for livestock.) And the same three foods that underpinned the world's earliest civilizations are still the foundations of human existence: Wheat, rice, and maize continue to provide the bulk of the calories consumed by the human race. The vast majority of the remaining calories are derived from domesticated plants and animals. Only a small proportion of the food consumed by humans today comes from wild food sources: fish, shellfish, and a sprinkling of wild berries, nuts, mushrooms, and so on.

Accordingly, almost none of the food we eat today can truly be described as natural. Nearly all of it is the result of selective breeding—unwitting at first, but then more deliberate and careful as farmers propagated the most valuable characteristics found in the wild to create new, domesticated mutants better suited to human needs. Corn, cows, and chickens as we know them do not occur in nature, and they would not exist today without human intervention. Even orange carrots are man-made. Carrots were originally white and purple, and the sweeter orange variety was created by Dutch horticulturalists in the sixteenth century as a tribute to William I, Prince of Orange. An attempt by a British supermarket to reintroduce the traditional purple variety in 2002 failed, because shoppers preferred the selectively bred orange sort.

All domesticated plants and animals are man-made technologies. What is more, almost all of the domesticated plants and animals on which we now rely date back to ancient times. Most of them had been domesticated by 2000 B.C., and very few have been added since. Of the fourteen large animals to have been domesticated only one, the reindeer, was domesticated in the past thousand years; and it is of marginal value (tasty though it is). The same goes for plants: Blueberries, strawberries, cranberries, kiwis, macadamia nuts, pecans, and cashews have all been domesticated relatively recently, but none is a significant foodstuff.

Only aquatic species have been domesticated in significant quantities in the past century. In short, early farmers managed to domesticate most of the plants and animals worth bothering with many thousands of years ago. That may explain why domesticated plants and animals are so widely assumed to be natural, and why contemporary efforts to refine them further using modern genetic-engineering techniques attract such criticism and provoke such fear. Yet such genetic engineering is arguably just the latest twist in a field of technology that dates back more than ten thousand years. Herbicide-tolerant maize does not occur in nature, it is true—but nor does any other kind of maize.

The simple truth is that farming is profoundly unnatural. It has done more to change the world, and has had a greater impact on the environment, than any other human activity. It has led to widespread deforestation, environmental destruction, the displacement of "natural" wildlife, and the transplanting of plants and animals thousands of miles from their original habitats. It involves the genetic modification of plants and animals to create monstrous mutants that do not exist in nature and often cannot survive without human intervention. It overturned the hunter-gatherer way of life that had defined human existence for tens of thousands of years, prompting humans to exchange a varied, leisurely existence of hunting-and-gathering for lives of drudgery and toil. Agriculture would surely not be allowed if it were invented today. And yet, for all its faults, it is the basis of civilization as we know it. Domesticated plants and animals form the very foundations of the modern world.

PART II

Food and Social
Structure

3

FOOD, WEALTH, AND POWER

Wealth is hard to come by, but poverty is always at hand.

—MESOPOTAMIAN PROVERB, 2000 B.C.

TINKER, TAILOR, SOLDIER, SAILOR

The Standard Professions List is a document from the dawn of civilization, inscribed in the characteristic wedge-shaped indentations of cuneiform script on small clay tablets. The earliest versions, dating from around 3200 B.C., were found in the city of Uruk (modern-day Erech) in Mesopotamia, the region where writing and cities first emerged. Many copies exist, since it was a standard text that was used to teach scribes. The list consists of 129 professions, always written in the same order, with the most important at the top. Entries include "supreme judge," "mayor," "sage," "courtier," and "overseer of the messengers," though the meaning of many entries is unknown. The list illustrates that the population of Uruk, probably the biggest city on earth at the time, was stratified into different specialist professions, some more important than others. This was a big change from the villages of farmers that had emerged in the region around five thousand years earlier. Food lay at the root of this transformation.

The switch from small, egalitarian villages to big, socially stratified cities was made possible by an intensification of agriculture in which part of the population produced more food than was needed for its own subsistence. This surplus food could then be used to sustain others—so not everyone had to be a farmer anymore. In Uruk,

31

only around 80 percent of the population were farmers. They tended fields that surrounded the city in a vast circle, ten miles in radius. Their surplus production was appropriated by a ruling elite at the top of society, which redistributed some of it and consumed the rest. This stratification of society, made possible by agricultural food surpluses, happened not just in Mesopotamia but in every part of the world where farming was adopted. It was the second important way in which food helped to transform the nature of human existence. With agriculture, people settled down; with intensification, they divided into rich and poor, rulers and farmers.

The idea that people have different jobs or professions, and that some are richer than others, is taken for granted today. But for most of human existence this was not the case. Most hunter-gatherers, and then early farmers, were of comparable wealth and spent their days doing the same things as the other people in the same community. We are used to thinking of food as something that brings people together, either literally around the table at a social gathering, or metaphorically through a shared regional or cultural cuisine. But food can also divide and separate. In the ancient world, food was wealth, and control of food was power.

As with the adoption of farming, the changes in food production and the associated transformation of social structures took place simultaneously and were intertwined. A ruling elite did not suddenly appear and demand that everyone else work harder in the fields; nor did greater productivity produce a sudden surplus to be fought over, with the winner crowned king. Instead, the abandonment of the hunter-gatherer lifestyle meant that previous constraints on individuals' ability to amass goods and cultivate prestige, both of which are frowned upon by hunter-gatherers, no longer applied. Even so, the emergence of more complex societies took some time: In Mesopotamia, the shift from simple villages to complex cities took five millennia, and it also took thousands of years in China and the Americas.

Control of food was power because food literally kept everything going, by feeding humans and animals. Appropriating the food surplus from farmers gave ruling elites the means to sustain full-time scribes, soldiers, and specialist craft workers. It also meant that a certain proportion of the population could be pressed into service on construction projects, since the farmers who remained on the land would provide enough food for everyone. So a store of surplus food conferred upon its owner the power to do all kinds of new things: wage wars, build temples and pyramids, and support the production of elaborate craft items by specialist sculptors, weavers, and metalworkers. But to understand the origins of food power it is necessary to start by examining the structure of hunter-gatherer societies, and to ask why people had previously regarded the accumulation of food and power to be so dangerous and destabilizing—and why this changed.

ANCIENT EGALITARIANS

Hunter-gatherers may only have had to spend two days a week foraging for things to eat, but their lives were nonetheless ruled by food. Bands of hunter-gatherers have to be nomadic, moving every few weeks once the food resources within range of each temporary camp start to become depleted. Every time the group moves, it has to take all of its possessions with it. The need to carry everything limits individuals' ability to accumulate material goods. An inventory by modern anthropologists of a family of African hunter-gatherers, for example, found that they collectively owned a knife, a spear, bow and arrows, a wrist guard, a net, baskets, an adze, a whistle, a flute, castanets, a comb, a belt, a hammer, and a hat. Few families in the developed world could list all their possessions in a single short sentence. These items were, furthermore, collectively owned and freely shared. Rather than having everyone carry his or her own knife or spear, it makes more sense to share such items, since some people can then carry other things, such as nets or bows. Bands in which items were shared would have been more flexible and more likely to survive than bands in which items were jealously guarded by individuals. So bands in which there was social pressure to share things would have proliferated.

The obligation to share also extended to food. Modern hunter-gatherers often have a rule that anyone who brings food back to the camp has to share it with anyone else who asks. This rule provides insurance against food shortages, for not everyone can be sure to find enough food on a given day, and even the best hunters can only expect to kill an animal every few days. If everyone is selfish and insists on keeping their own food to themselves, most people will be hungry most of the time. Sharing ensures that the food supply is evened out and most people have enough to eat most of the time. Ethnographic evidence from modern hunter-gatherers shows that some groups have even more elaborate rules to enforce sharing. In some cases a hunter is not even allowed to help himself to food from his own kill (though a family member will ensure that some food is passed to him indirectly). Similarly, trying to claim a patch of land, and its associated food resources, is not allowed. Such rules ensure that the risks and rewards of hunting and gathering are shared throughout the group. Historically, bands that practiced food sharing were more likely to survive than those that did not: Competition for resources tends to encourage overexploitation, and ownership disputes would have caused bands to fragment. Once again, food sharing predominated because it conferred clear advantages upon bands that adopted it.

All of this meant that hunter-gatherers did not try to accumulate status goods to enhance their personal prestige. Why bother, since such goods would have had to have been shared with others? It is not until the advent of agriculture that the first indications of wealth or private ownership appear. As one anthropologist noted, having observed hunter-gatherers in Africa:

A Bushman will go to any lengths to avoid making other Bushmen jealous of him, and for this reason the few possessions the Bushmen have are constantly circling among members of their groups. No one cares to keep a particularly good knife long, even though he may want it desperately, because he will become the object of envy; as he sits by himself polishing a fine edge on the blade he will hear the soft voices of the other men in his band saying: "Look at him there, admiring his knife while we have nothing." Soon somebody will ask him for his knife, for everybody would like to have it, and he will give it away. Their culture insists that they share with each other, and it has never happened that a Bushman failed to share objects, food or water with other members of his band, for without very rigid co-operation Bushmen could not survive the famines and droughts that the Kalahari offers them.`

Hunter-gatherers are also suspicious of self-promotion and attempts to create obligation. The !Kung Bushmen, for example, believe that the ideal hunter should be modest and understated. After returning from the hunt he must downplay his achievements, even if he has killed a large animal. When the men go to fetch the kill, they then express their disappointment at its size: "What, you made us come all this way for this bag of bones?" The hunter is expected to play along, and not to be offended. All of this is intended to prevent the hunter from regarding himself as superior. As one !Kung Bushman explained to a visiting ethnographer: "When a young man kills much meat, he comes to think of himself as a chief or a big man, and he thinks of the rest of us as his servants or inferiors. We can't accept this. So we always speak of his meat as worthless. In this way we cool his heart and make him gentle."

To further complicate matters, the !Kung have a tradition that the meat from a kill belongs to the owner of the arrow that killed it, rather than the hunter who fired it. (If two or more arrows bring down the kill, the meat belongs to the owner of the first arrow.) Since the men routinely exchange arrows, this makes grandstanding by individual hunters even less likely. Particularly skilled hunters are thus prevented from cultivating prestige for themselves by conferring large amounts of food on others and so creating an obligation.

Quite the opposite, in fact: When a hunter has had a run of good luck and produced a lot of food, he might stop hunting for a few weeks in order to give others the chance to do well, and so avoid any possibility of resentment. Taking a few weeks off also means the hunter can allow others to provide him with food, so that there is no question of an outstanding obligation to him.

Richard Borshay Lee, a Canadian anthropologist who lived with a group of !Kung on several research trips during the 1960s, ran afoul of these rules when he tried to thank his hosts by holding a feast for them. He bought a large, plump ox for the purpose and was surprised when the Bushmen began to ridicule him for having chosen an animal that was too old, too thin, or would be too tough to eat. In the event, however, the meat from the ox turned out to be tasty and tender after all. So why had the Bushmen been so critical? "The !Kung are a fiercely egalitarian people and have a low tolerance for arrogance, stinginess and aloofness among their own people," Lee concluded. "When they see signs of such behaviour among their fellows, they have a range of humility-enforcing devices to bring people back into line." The !Kung, like other hunter-gatherers, regard lavish gifts as an attempt to exert control over others, curry political support, or raise one's own status, all of which run counter to their culture. Their strict egalitarianism can be regarded as a "social technology" developed to ensure social harmony and a reliable supply of food for everyone.

Food determines the structure of hunter-gatherer society in other ways, too. The size of hunter-gatherer bands, for example, depends on the availability of food resources within walking distance of the camp. Too large a band depletes the surrounding area more quickly, which makes it necessary to move the camp more often and means the band needs a larger territory. As a result, band sizes vary between six to twelve people in areas where food is scarce and twenty-five to fifty people in areas with more abundant resources. The bands consist of one or more extended families, and because of intermarriage most members of the band are related to each other. Bands generally do not have leaders, though some people may have particular roles in addition to the traditional male and female tasks of hunting and gathering, respectively, such as healing, making weapons, or negotiating with other bands. But there are no full-time specialists, and these particular skills do not confer a higher social status.

Hunter-gatherer bands maintain alliances with other bands, to provide both marriage partners and further insurance against food shortages. In the event of a shortage one band can then visit another to which it is related by marriage and share some of its food. Intergroup sharing in the form of large feasts also takes place at times of seasonal food overabundance. Such feasts appear to be universal among hunter-gatherers and provide an opportunity to arrange marriages, perform social rituals, sing, and dance. Food thus binds hunter-gatherer societies together, forging links both within bands and between bands.

That said, it is important not to over-romanticize the hunter-gatherer lifestyle. The "discovery" of surviving hunter-gatherer bands by Europeans in the eighteenth century led to the creation of the idealized portrait of the "noble savage" living in an unspoiled Eden. When Karl Marx and Friedrich Engels developed the doctrine of communism in the nineteenth century, they were inspired in part by the "primitive communism" of hunter-gatherer societies described by Lewis H. Morgan, an American anthropologist who studied Native American societies. But even though the hunter-gatherer life was more leisurely and egalitarian than most people's lives are today, it was not always idyllic. Infanticide was used as a means of population control, and there was routine and widespread conflict between hunter-gatherer bands, with evidence of violent death and even cannibalism in some cases. The notion that hunter-gatherers lived in a perfect and peaceful world is beguiling but wrong. Even so it is clear that the structure of hunter-gatherer society, which was chiefly determined by the nature of the food supply, was strikingly different from that of modern societies. So when people took up farming, and the nature of the food supply was transformed, everything changed.

The Emergence of the "Big Man"

As people began to settle down and hunting and gathering shaded into farming, the first villages were still broadly egalitarian communities. Archaeological evidence shows that the earliest such villages, typically inhabited by no more than one hundred people, were made up of huts or houses of similar shape and size. But settlement and agriculture changed the rules that had previously discouraged people from pursuing wealth and status. The social mechanisms that had been developed to suppress man's inherent tendencies toward hierarchical organization (clearly visible in apes and many other animal species) began to erode. Once you are no longer moving around, it starts to become possible to amass surplus food and other goods. The first signs of social differentiation begin to appear: villages in which some dwellings are larger than others and contain prestige items such as rare shells or ornate carved items, and burial grounds in which some graves contain valuable grave goods and others from the same period do not. All of this implies that the concept of private property quickly became accepted—there is no point in owning status goods if you have to share them—and a social hierarchy started to emerge in which some people were richer than others.

In some places, this process began even before the advent of agriculture, as hunter-gatherers in particularly food-rich areas settled down in permanent villages. But it became widespread with the adoption of farming. Early agricultural villages in China's Hupei basin on the Upper Yangtze River, in the region where rice was domesticated around 4000 B.C., provide a good example. Of 208 graves excavated, some contained elaborate grave goods, while others contained nothing more than the bodies of the dead. Similarly, 128 graves dating from around 5500 B.C. at Tell es-Sawwan, in what is now northern Iraq, show a clear variation in grave goods. Some graves contain carved alabaster, beads made from exotic stones, or pottery, but others contain no grave goods at all. In each case the pattern is the same:

The adoption of agriculture leads to social stratification, subtle at first but then increasingly pronounced.

It is easy to see how variations in different families' agricultural productivity, and the ability to store certain foods (notably dried cereal grains), would make people more inclined to assert ownership over their produce. And since a storable food surplus can be traded for other items, it is equivalent to wealth. But a village in which some inhabitants manage to accumulate more food and trinkets than others is still a far cry from the elaborate social hierarchies of the first cities, in which the ruling elites appropriated the surplus by right and then distributed the portion of it they did not consume themselves. How did these powerful leaders emerge, and how did they end up in control of the agricultural surplus?

An important step along the road from an egalitarian village to a stratified city seems to be the emergence of "big men" who win control of the flow of surplus food and other goods and so amass a group of dependents or followers. Perhaps surprisingly, the big man's means of persuasion is not the threat of violence, but his abundant generosity. By bestowing gifts on others he places them in his debt, and they must reciprocate with more generous gifts in the future. Such gifts most often take the form of food. To get the ball rolling, a big man might persuade his family to produce surplus food, which he then gives to others; he subsequently receives more food in return, which he can then distribute among his family and give to others, thus conferring further obligations. This process can still be observed today, since big-man cultures still exist in some parts of the world.

In Melanesia, a big man might take several wives in order to increase the resources at his disposal to give away: one wife to garden and produce food, one to collect wood, another to catch fish. He then deploys these resources carefully, putting other people in his debt, so that they must repay him with even more, which he passes on to others, thus securing an even greater obligation. This process encourages intensification of food production, and eventually it culminates in big feasts as the big man tries to "build his name." He invites people from outside his existing circle, and even from other villages, thus placing them in his debt as well and bringing them into his sphere of influence. In this way, the big man establishes himself as an influential and powerful member of the community. Rivalry between big men accelerates the process, as they vie to hold the biggest feasts and amass the most credit.

Does this mean big men are rich and lazy? Far from it. For a big man, wealth is not something to sit on, but something that is only useful if it is given away. In some cases big men may even end up being poorer than their followers. In North Alaskan Eskimo groups, for example, the most respected whaling captains are responsible for trading with inland caribou hunters, and therefore end up controlling the distribution and circulation of valuables within their group. But since they must give away everything they receive, and cannot refuse a request for help, they are often materially worse off than their followers. Big men must work hard, too. According to one observer in Melanesia, the big man "has to work harder than anyone else to keep up his stocks of food. The aspirant for honours cannot rest on his laurels but must go on holding large feasts and piling up credits. It is acknowledged that he has to toil early and late."

All of this actually serves a useful purpose within the group or village: The big man acts as a clearinghouse for surplus food and other goods and can determine how best to distribute them. If a family produces extra food, it can give the surplus to a big man with the expectation of being able to call in the favor at a later date—when a tool needs replacing, perhaps, or food runs short. A successful big man thus integrates and coordinates the economy of the community, and he emerges as its leader. But he has no power to coerce his followers. Maintaining his position depends on being able to provide for the group and govern redistribution. Among the Nambikwara of Brazil, for example, if the leader of the group is not generous enough and fails to provide for his followers, they will leave and join a different group. Within Melanesian groups, leaders who fail to deliver or who try to keep too much of the surplus for themselves may be deposed or even murdered. In such a situation the big man is still far more of a manager than a king.

From Chiefdoms to Civilizations

So how does the big man, whose position depends on generosity and sharing, develop into the powerful chief of a group of villages, or chiefdom, and then the king at the top of a ruling elite? Not surprisingly, as with the origins of agriculture and the spread of farming, the mechanism is unclear and there are many competing theories. And once again it is likely that no single theory provides the answer, and some explanations are more valid in some parts of the world than others. Yet by looking at several such theories it is possible to get an idea of how chiefdoms, and then civilizations, might have emerged. In each case, the emergence of social stratification is tightly bound up with the production of food. More elaborate forms of social organization make possible greater agricultural productivity, and a larger food surplus can support more elaborate forms of social organization. But how does the process start?

One theory contends that a big man or leader can become more powerful by coordinating agricultural activity, particularly irrigation. Farming yields can vary widely, but by leveling land and building irrigation canals and levee systems—all of which is only possible with a certain amount of social organization—it is possible to reduce these variations. This would increase the village's agricultural productivity, and would have other effects too. Members of the community would be less inclined to leave once they had invested in irrigation systems and had come to rely on them; control of the irrigation system would confer power on the leader, since anyone who fell out of favor might have his water supply reduced; the irrigation system might also need to be defended, using full-time soldiers funded by the food surplus and placed under the leader's control.

What starts off as a community farming project, in short, could have the effect of greatly increasing the leader's power. With his followers more dependent on him and a private guard to protect him, he would then be able to start retaining more of the surplus for his own use, to feed his household, provision soldiers, and so on. Irrigation systems are certainly a common denominator of many early civilizations, from Mesopotamia to Peru. They are found in chiefdoms, too, in places such as Hawaii and southwestern North America. But some chiefdoms that relied on irrigation did not go on to become any more complex or sharply stratified; and some elaborate irrigation schemes seem to be the consequences of greater organization rather than its cause. So evidently there is more to the emergence of complex civilizations than irrigation, though it seems to have played a role in some cases.

Another theory suggests that the communal storage of agricultural surpluses might provide the leader with an opportunity to establish greater control over his followers. Villagers hand over surplus grain to the big man in anticipation of reciprocal gifts at a later date, prompting him to organize the construction of a granary. Once built and provisioned, it provides the big man with the "working capital" to do other things. He can fund full-time craft specialists and organize agricultural works using the surplus, on the basis that such investments produce a positive return that can be put back into the granary. Increasingly elaborate public-works projects then legitimize the leader's position and require a growing number of administrators, who emerge as the ruling elite. According to this view, there is a natural progression from reciprocal sharing organized by a big man to redistribution overseen by a powerful chief.

In the Near East, large central buildings started to appear within villages after around 6000 B.C., but it is unclear whether they were shared granaries, feasting halls, religious buildings, or chiefs' houses. They may well have served several of these functions: A feasting hall built to impress the neighboring village might have been the logical place to store food, and a granary would have been an obvious place to perform fertility rituals to ensure a good harvest. There is evidence from Hawaii that what were originally public areas built for feasting were later walled off, with access restricted to a select group of high rank. So temples and palaces could have started out as communal storehouses or feasting halls.

A third suggestion is that competition for agricultural land led to warfare between communities in areas where such land was environmentally circumscribed. In Peru, for example, seventy-eight rivers run from the Andes mountains to the coast through fifty miles of extremely dry desert. Agriculture is possible near the rivers, but all the suitable farming areas are hemmed in by desert, mountains, and oceans. In Egypt, farming is possible on a narrow ribbon of fertile land along the Nile, but not in the desert beyond. And on the alluvial plains of Mesopotamia, only areas near the Tigris and Euphrates rivers are suitable for farming. To start with, such areas would have been lightly populated by a few farmers. As the population of farmers expanded (since sedentism and farming enable population growth beyond hunter-gatherer levels) new villages would have been established. Once all the available farming land was being used, farmers intensified production, extracting more food from a given area using elaborate terraces and irrigation systems.

But eventually they reached the limit of agricultural productivity, at which point the villages began to attack each other. When one village defeated another it then appropriated the defeated village's land or forced its people to hand over a proportion of their harvest every year. In this way, the strongest village within an area emerged as the ruling class, and the weaker villages had to hand over their surplus production, thereby establishing a system in which the poor farmed for the rich. This all sounds plausible, but there is no evidence that people reached the limit of agricultural productivity in any of the places where stratified societies first emerged. In the event of a drought or a bad harvest, however, it is possible to imagine villages with food reserves coming under attack from neighboring villages where the food had run out.

A more general view that encompasses all of these theories is the idea that more complex societies (that is, those with strong leadership and a clear social hierarchy) will be more productive, more resilient, better able to survive hardship, and better at defending themselves. Villages in which strong leaders emerge would then outcompete other, less well organized villages nearby, and would be more attractive places to live, at least for those who do not mind submitting to the leader's authority. The emergence of strong leaders is often assumed to be dependent on coercion, but people might initially have regarded the need to hand over some or all of their surplus production to the leader as a price worth paying if the benefits they received in return—working irrigation systems, greater security, performance of religious rites to maintain soil fertility, mediation in disputes—were deemed to be of sufficient value. But the leader would then have been in a position to keep more and more of the surplus for his own use. Once you have settled down and invested labor in a house, fields, and irrigation systems, you have a reason to stay put even if the leader starts to put on airs and graces, claims he is descended from a god, and so on.

How can we tell what happened? The archaeological evidence shows the process of social stratification happening around the world in much the same way, culminating in the emergence of broadly comparable Bronze Age civilizations in different parts of the world, but at different times: starting in Egypt and Mesopotamia around 3500 B.C.; during the Shang dynasty in northern China around 1400 B.C.; with the rise of Maya civilization in southern Mexico from around 300 A.D.; and in South America around the same time, leading to the establishment of the Inca Empire in the 15th century A.D.

The trouble is that the archaeological evidence does not reveal much about the mechanism of stratification. The first signs of change are usually greater variations in grave goods and the emergence of more elaborate regional pottery styles, which appear around 5500 B.C. in Mesopotamia, 2300 B.C. in northern China, and 900 B.C. in the Americas. Such pottery suggests some degree of specialization, and possibly the emergence of elites capable of supporting full-time craft workers. Huge numbers of pottery bowls made in standard sizes appear in Mesopotamia around 3500 B.C., which suggests that their manufacture had been placed under centralized control and that standard measures of grain and other commodities were used when paying taxes and distributing rations.

In northern China, settlements from the Longshan period (3000–2000 B.C.) start to have large walls, and weapons such as spears and clubs become more widespread. In Mesopotamia, L-shaped entrances to buildings, caches of stones for use in slingshots, and defensive earthworks appear. All are suggestive of organization for the purpose of defense. Just as telling are the first steps toward writing, in the form of tokens and seals used for administration in Western Asia and symbols written on bones by specialist fortune-tellers in northern China. Ever-larger settlements, as villages grow into towns, indicate greater political organization for the simple reason that without some accepted authority to adjudicate when disputes arise, villages seem unable to grow beyond a certain size.

By the start of the Shang dynasty in China around 1850 B.C. there are dedicated craft workshops, and some settlements have some kinds of workshop but not others, suggesting deliberate local specialization. The ability to work bronze in the Near East and China and gold in South America is another sign of craft specialization, and the presence of fine metalwork in grave goods signals stratification, in some cases to an extraordinary degree. In the "royal" tombs of the Mesopotamian city of Ur, dating from around 2500 B.C., the dead were entombed with gold, silver, and jewel-encrusted items. They were also accompanied by dozens of sacrificed servants, musicians, and bodyguards, and even by oxen to draw their chariots. These tombs, and similar examples in China, provide striking and gruesome evidence of social stratification.

By the time the first cities appear, with their specialist craftsmen organized into districts, and monumental buildings such as temples and pyramids, there is no question that social stratification has occurred. Indeed, there is direct written evidence of it. In China, documents detail a complex hierarchy of nobles, each with his own territory, under a king. In Mesopotamia's city-states, clay tablets record taxes paid, commodities produced, and rations issued; there are also membership lists for specialist guilds, from brewers to snake charmers. In Egypt, the Overseer of All the Works of the King in the Fourth Dynasty (the period in which the pyramids were built) had a large staff of officials and scribes who scheduled, fed, and organized large numbers of full-time masons and even more numerous rotating teams of construction workers. This involved a mountain of ration lists and timetables.

The appearance of monumental architecture, many examples of which are still standing today around the world, undoubtedly provides the most direct and enduring evidence of the social stratification of the first civilizations. Such large-scale building works can only be carried out under an efficient system of administration, with

A Mesopotamian depiction of a city, with different kinds of workers overseen by a king.

a system to store surplus food and issue it as rations to building workers and an ideology to convince people that the construction project is worthwhile—in short, by a hierarchical society ruled by an all-powerful king. The defining characteristic of such tombs, temples, and palaces is that they are far bigger and more elaborate than they need to be. Such buildings are statements of power, and as societies become more stratified, these buildings become more prominent.

The pyramids of Egypt, the ziggurats of Mesopotamia, and the stepped temples of central and southern Mexico were made possible by agricultural food surpluses and the associated increase in social complexity. Hunter-gatherers would not have dreamed of building them, and even if they had, they lacked the means—the wealth in the form of surplus food, and the necessary organizational structures—to do so. These great edifices stand as monuments to the rise of the first civilizations, but also to the emergence of a new and unprecedented degree of inequality and social stratification that has persisted ever since.

4

FOLLOW THE FOOD

He rained down manna also upon them for to eat:
and gave them food from heaven.

—PSALM 78, VERSE 25

FOOD AS A TRACER FOR POWER STRUCTURES

Just before sunrise on a May morning, more than six hundred richly dressed Inca youths lined up in two parallel rows in a sacred field, surrounded by swaying stalks of maize. As the first glimmers of the sun appeared, they began to sing, quietly at first but with gathering intensity as the sun rose into the sky. Their song was a military victory chant, or *haylli*. The singing built in volume throughout the morning, reaching a climax at noon. It then grew gradually quieter during the afternoon and ended when the sun set. In the twilight the young men, who were all newly initiated sons of Inca nobles, began to harvest the crop. This scene, repeated every year, was just one of several maize-related Inca customs that demonstrated and reinforced the privileged status of the ruling elite.

Another example was the maize-planting ceremony that took place in August. When the sun set between two great pillars on the hill of Picchu, as seen from the center of Cuzco, the Inca capital, it was time for the king to initiate the growing season. He did so by plowing and planting one of several sacred fields that could only be tilled and worked by members of the nobility. According to one eyewitness account: "At sowing time, the king himself went and ploughed a little . . . the day when the Inca went to do this was a solemn festival of all the lords of Cuzco. They made great sacrifices to this flat place, especially of silver, gold and children." The plowing was then carried on by Inca nobles, but only after the king had started the process. "If the Inca had not done this, no Indian would dare to break the earth, nor did they believe it would produce if the Inca did not break it first," noted another observer. Further sacrifices of llamas and guinea pigs were made as the maize planting began. In the middle of the field priestesses poured *chicha*, or maize beer, onto the soil around a white llama. These offerings were to protect the fields from frost, wind, and drought.

For the Incas, agriculture was closely linked to warfare: The earth was defeated, as if in battle, by the plow. So the harvest ceremony was carried out by young noblemen as part of their initiation as warriors, and they sang a haylli as they harvested the maize to celebrate their victory over the earth. At the beginning of the next growing season, only the ruling Inca had the power to defeat the earth and capture its reproductive energies to ensure the success of the agricultural cycle, so he had to break the ground first. This emphasized his power over his people: Without him, they would starve. The symbolic defeat of the earth was also a reenactment of the battle between the first Incas and the indigenous inhabitants of Cuzco, the Hualla, whom the Incas had defeated before planting the first corn. As the Incas saw it, they had triumphed over nature in two ways: by defeating the local savages and then by introducing agriculture. The ruling elite claimed to be the direct descendants of the winners of that original battle. The ceremonies highlighted this link, and hence the right of the elite to rule over the masses, while also suggesting that the hierarchical structure of society was part of an ancient natural order. The implication was that if the king and his nobles were overthrown, there would be nobody to make the crops grow.

Food-related activities of this kind were widely used to define and reinforce the privileged position of the elite in early civilizations. Food, or food-production capacity, was used to pay tax. Food was extracted as tribute after military victories. Food offerings and sacrifices were used to maintain the stability of the universe and ensure the continuation of the agricultural cycle. Formal handouts of food, as rations and wages and at feasts and festivals, also emphasized how food, and hence power, was distributed. In the modern world, you follow the money to determine where power lies. In the ancient world it is food that reveals power structures. To illuminate the organization of the first civilizations, you must follow the food.

FOOD AS CURRENCY

Food was used within early civilizations as a form of currency, in barter transactions, and to pay wages and taxes. Food was passed upward from the farmers to the ruling elite in various ways and then redistributed as wages and rations to support the elite's activities: building, administration, warfare, and so on. The principle that some or all of the agricultural surplus had to be handed over is common to all early civilizations, since the appropriation of the surplus had been central to their emergence in the first place. There were many different schemes. But in each case the structure of society—who people worked for, where their sustenance came from, and where their loyalties lay—was defined by food.

In Egypt and Mesopotamia, tax was paid both directly in the form of food and indirectly in the form of agricultural labor. Most Egyptian farmers did not own their own land but rented it from landowners, who claimed a fraction of the resulting harvest. The state owned a lot of land, so this produced a lot of food income. Other land belonged to officials, temples, nobles, and the pharaoh himself, and this too was rented to farmers in return for a share of their harvest, with a fraction of that rent going as tax to the state. The rent charged and tax levied depended on the agricultural potential of the land, given its proximity to wells and canals and the level of each year's Nile flood.

The Hekanakhte Papers, a set of letters dating from around 1950 B.C. written by a priest to his family while he was away from his estate, give details of this system in action, while also providing a rare glimpse of everyday life in Ancient Egypt. Hekanakhte seems to have been in charge of land belonging to a temple, and in his letters he advises his family about which bits of land to cultivate and how much each can be expected to yield, how many sacks of barley to charge when renting land to other farmers, and how many sacks of barley to pay the laborers on the estate. Evidently times are bad and food is scarce, and Hekanakhte reminds his family that they are eating better than most people. There is a quarrel over a handmaiden named Senen, and much indulgence is shown to a spoiled young man named Snofru. Debts and rents are collected in barley and wheat, and in some cases jars of oil are accepted as payment instead: one jar of oil is worth two sacks of barley, or three of wheat.

Tax, like rent, was also paid in the form of food, and tax collectors took the resulting goods to regional administrative centers, where they were redistributed as pay to government officials, craft workers, and farmers seconded to work for the state as corvée laborers. Such workers built and maintained irrigation systems, constructed tombs and pyramids, worked in mines, and performed military service. During a stint of corvée work, which might last for several months, laborers were fed, housed, and clothed by the state. It was corvée workers who built the pyramids; surviving ration lists show that they received daily portions of bread and beer, supplemented with onions and fish. A similar scheme prevailed in Mesopotamia, where land was owned by wealthy families, temples, city councils, or the palace. Farmers handed over a fraction of their harvest to rent land, and the king levied taxes on non-palace fields. In this way most of the surplus went to the king, the temples belonging to various gods, and landowners. As in Egypt, corvée labor was used in large construction projects.

In some cultures, however, taxes were paid solely in the form of labor. In Shang China, rural clans worked their own communally held fields, but they also cultivated special fields, the produce from which went to the king, to rural governors, or to other officials. Similarly, Inca farming families cultivated their own fields and those belonging to their clan, or *ayllu*. Produce from the ayllu's fields supported the local chief and the cult of the local god. Farmers also spent part of their time working on state-owned fields and on those belonging to temples of more important gods. This scheme arose from a deal struck when ayllu, which were previously autonomous communities, were incorporated into the Inca kingdom: The clans were allowed to keep their own land and its produce, provided they supplied labor to work state-owned fields in return. This meant that the Inca king was not given any food as tax by his subjects, which would have placed him in their debt; instead, they worked his land and he took the produce, which was transported to regional storehouses. Inca farmers also had to carry out corvée work from time to time, doing construction work, mining, or military service. All this was recorded using a system of colored, knotted strings called *quipus*.

Aztec society was divided into landholding groups called *calpullis*. Unlike Inca ayllu, all the members of which were equals under the chief, calpullis were overseen by a few high-ranking families who belonged to the Aztec nobility. Each family cultivated both its own fields and shared fields, the produce from which supported the calpulli's nobles, temples, teachers, and soldiers. Calpullis also had to provide a certain amount of tax and corvée labor to the Aztec state. In addition, the king, state institutions, and important nobles and warriors owned their own land, which was worked by landless farmers who were given just enough food to subsist on. The rest of the produce from this land went directly to its owners.

Food also flowed from subject states in the form of tribute, extracted by dominant states and city-states from the weaker neighbors under threat of military force, usually after a military defeat. Following the defeat of one city-state by another in Mesopotamia, for example, the losing city would be looted and would also have to pay regular tribute to the winning city. Sargon of Akkad, who conquered the city-states of Mesopotamia around 2300 B.C. and unified them into an empire, demanded vast amounts of tribute from each city: Inscriptions speak of entire warehouses of grain being paid. As well as emphasizing his superiority, this kept the subject cities weak and Sargon's capital strong. It also allowed him to support a huge staff: He boasted of feeding 5,400 men every day. By redistributing tribute among their followers, rulers could reinforce their leadership and maintain support for further military campaigns.

Perhaps the best example of tribute collection is that of the Aztec "triple alliance" between Tenochtitlan, Texcoco, and Tlacopan. These three city-states collected tribute from the whole of central Mexico. Nearby subject states in and around the Valley of Mexico had to supply huge quantities of food: Every day the chief of Texcoco received enough maize, beans, squashes, chiles, tomatoes, and salt to feed more than two thousand people. More distant states supplied cotton, cloth, precious metals, exotic birds, and manufactured items. The level of tribute paid depended on each state's distance from the three capitals (the alliance's control over those farther away was weaker, so it demanded less in tribute from them) and on whether the state put up a fight or not before submitting to alliance rule (states that gave in without a fight paid less). The constant flow of food and other goods toward the capital meant there was no doubt where the power lay. Aztec rulers used this tribute to pay officials, provision the army, and support public works. Tribute handed out to the nobility reinforced the ruler's position and simultaneously weakened the rulers of subordinate states, who ended up with less to distribute among their own followers: less food meant less power.

FEEDING THE GODS

As systems of social organization became more elaborate, so too did the religious practices that provided cosmological justification for the elite's right to levy all these taxes. Religious beliefs and traditions varied widely among the world's first civilizations, but in many cases there was a clear congruence between the payment of taxes by the masses to the elite and the "payment" of sacrifices and offerings by the elite to the gods. Such offerings were believed to return energy to its divine source, so that the source could continue to animate nature and supply humans with food. Rather than being so powerful that they could exist without humanity's support, the gods were thought to be dependent on humans, and humans were thought to be dependent in turn on the gods. An Egyptian text from around 2070 B.C. refers to humans as the creator god's "cattle," for example, implying that the god both looked after humans and depended upon them for his own sustenance. Similarly, many cultures believed that the gods had created mankind to provide spiritual nourishment in the form of sacrifices and prayers. In return, the gods provided physical nourishment for humans by making plants and animals grow. Sacrifices were regarded as an essential means of maintaining this cycle.

Some Mesoamerican cultures believed that the gods even sacrificed themselves or each other from time to time to ensure the continued existence of the universe and survival of mankind. The Maya, for example, believed that maize was the flesh of the gods containing divine power, and at harvest time the gods were, in effect, sacrificing themselves to sustain humanity. This divine power passed into humans as they ate, and was particularly concentrated in their blood. Human sacrifices in which blood was spilled were a way to repay this debt and return the divine power to the gods. Food and incense were provided as offerings as well, but human sacrifices were thought to be most important of all.

The Aztecs also regarded human sacrifices as a way to repay energy owed to the gods. The Earth Mother was nourished by human blood, they believed, and the crops would only grow if she was given enough of it. It was supposedly an honor to be sacrificed, but even so victims seem not to have belonged to the ruling elite. Instead, they were mostly criminals, prisoners of war, and children. Human flesh and blood were thought to be made from maize, so these sacrifices sustained the cosmic cycle: Maize became blood, and blood was then transformed back into maize. Sacrificial victims were referred to as "tortillas for the gods." The Incas also thought sacrifice was necessary to nourish the gods. They offered llamas, guinea pigs, birds, cooked vegetables, fermented drinks, cocoa, gold, silver, and elaborately woven cloth, which was burned to release the energy that had gone into weaving it. Food and alcoholic drinks made from maize were thought to be particularly favored by the gods. But most valued of all were human sacrifices. After subjugating a new region, the Incas sacrificed its most beautiful people.

In Egyptian temples, animals were killed and their flesh was presented to images of the gods. The gods were believed to inhabit the images three times a day in order to consume the life force from the offerings, which they needed to replenish the energy they expended to keep the universe going. Food offerings were also required to maintain the life force of dead humans, who had become gods. So offerings were frequently made to dead pharaohs, and tombs were filled with jars of food to sustain the dead in the afterlife. Similarly, in Shang China both gods and royal ancestors were offered grain, millet beer, animals (dogs, pigs, wild boars, sheep, and cattle), and human sacrifices, most of them prisoners of war. The gods were thought to drink the blood of the slaughtered victims. But the most elaborate offerings were made to the ancestors of Shang kings, who depended on these sacrifices as food. If their ancestors were not sufficiently well fed, the Shang kings believed, they would punish their descendants with poor harvests, military defeats, and plagues.

The Mesopotamians thought humans had a duty to provide food and earthly residences for the gods, who were provided with two meals a day in their temples. The gods depended on this nourishment from humans: In the Mesopotamian version of the flood story, the gods destroy humanity and then regret their action when they grow hungry because of the lack of offerings. But one of their number, Enki, warns Utnapishtim (the Mesopotamian equivalent of the biblical Noah) of the coming flood and tells him to build an ark. When Utnapishtim emerges from his boat and offers a burnt sacrifice, the gods crowd around the smoke "like flies" because it is the first nourishment they have had in days. They then forgive Enki for allowing a few humans to survive. The Mesopotamians believed the gods could survive without humans, but only if they produced their own food—which is why they created humans to do it for them, and taught humans about agriculture.

In all these cases, sacrifices and offerings channel energy back to the supernatural realm as spiritual food to nourish gods and ancestors and ensure that they, in turn, continue to nourish mankind by keeping the agricultural cycle going. The presentation of sacrifices gave the elite a crucial intermediary role between the gods and the farming masses. By paying tax, the farmers in effect exchanged food for earthly order and stability, as the elite managed irrigation systems, organized military defenses, and so on. And by providing sacrifices to the gods, the elite in effect exchanged spiritual food for cosmic order, as the gods maintained the stability of the universe and the fertility of the soil.

That such similar religious ideologies arose in the earliest civilizations, separated as they were in time and space, is surely no coincidence. The notion that the gods depended on offerings from mankind for their survival was peculiar to these cultures, no doubt because it was very convenient for the members of their ruling elites. It legitimized the unequal distribution of wealth and power and provided an implicit warning that without the managerial activities of the elite, the world would come to an end. The farmers, their rulers, and the gods all depended on each other to ensure their survival; catastrophe would ensue if any of them deviated from their assigned roles. But just as the farmers had a moral imperative to provide food to the elite, the elite in turn had a duty to look after the people and keep them safe and healthy. There was, in short, a social compact between the farmers and their rulers (and, by extension, the gods): If we provide for you, you must provide for us. The result was that taxes paid in earthly food and sacrifices of spiritual food, all justified by religious ideology, reinforced the social and cultural order.

The Agricultural Origins of Inequality

In the modern world, the direct equation of food with wealth and power no longer holds. For people in agricultural societies, food functions as a store of value, a currency, and an indicator of wealth; it is what people toil all day to produce. But in modern urban societies, money performs these roles instead. Money is a more flexible form of wealth, easily stored and transferred, and it can be readily converted into food at a supermarket, corner shop, café, or restaurant. Food is only equivalent to wealth and power when it is scarce or expensive, as it was for most of recorded history. But by historical standards, food today is relatively abundant and cheap, at least in the developed world.

Yet food has not entirely lost its association with wealth. It would be strange if it had, given how far back the connection goes. Even in modern societies there are numerous echoes of food's once-central economic role, in both words and customs. In English a household's main earner is called the breadwinner, and money may be referred to as bread or dough. Shared meals are still a central form of social currency: The elaborate dinner party must be reciprocated with an equally lavish meal in return. Extravagant feasts are a popular way to demonstrate wealth and status and, in the business world, to remind people who is boss. And in many countries the poverty line is defined in relation to the income required to purchase a basic minimum of foodstuffs. Poverty is a lack of access to food; so wealth, by implication, means not having to worry about where your next meal is coming from.

A common feature of wealthy societies, however, is a feeling that an ancient connection with the land has been lost, and a desire to reestablish it. For the wealthiest Roman nobles, knowledge of agriculture and ownership of a large farm was a way to demonstrate that they had not forgotten their people's purported origins as humble farmers. Similarly, many centuries later in pre-revolutionary France,

Queen Marie-Antoinette had an idealized farm built on the grounds of the palace of Versailles, where she and her ladies-in-waiting would dress up as shepherdesses and milkmaids, and milk cows that had been painstakingly cleaned. Today, people in many wealthy parts of the world enjoy growing their own food in gardens or on allotments. In many cases they could easily afford to buy the resulting fruit and vegetables instead, but growing their own food provides a connection with the land, a gentle form of exercise, a supply of fresh produce, and an escape from the modern world. (Growing food without the use of chemicals is often particularly highly regarded in such circles.) In California, the richest part of the richest country in the world, it is the simple food of the Italian peasantry that is most highly venerated. A tourist village has even opened in India, near the technology hot spot of Bangalore, where the newly prosperous middle classes can go to experience a romanticized version of their forebears' existence as subsistence farmers. One of the privileges of wealth is the option to emulate the lifestyles of the rural poor.

Wealth tends to distance people from working on the land; indeed, not having to be a farmer is another way to define wealth. Today, the richest societies are those in which the proportion of income spent on food, and the fraction of the workforce involved in food production, are lowest. Farmers account for only around 1 percent of the population in rich countries such as the United States and Britain. In poor countries such as Rwanda, the proportion of the population involved in agriculture is still more than 80 percent—as it was in Uruk 5,500 years ago. In the developed world, most people have specialized jobs that do not relate to agriculture, and they would find it difficult to survive if they suddenly had to produce all their own food. The process of separation into different roles that began when people first took up farming, and abandoned the egalitarian hunter-gatherer lifestyle, has reached its logical conclusion.

That people in the developed world today generally have a specific job—lawyer or mechanic or doctor or bus driver—is a direct consequence of food surpluses resulting from a continuous increase in the productivity of farming over the past few thousand years. Another corollary of these burgeoning food surpluses was the division into rich and poor, powerful and weak. None of these distinctions can be found within a hunter-gatherer band, the social structure that defined mankind for most of its existence. Hunter-gatherers own few or no possessions, but that does not mean they are poor. Their "poverty" only becomes apparent when they are compared with members of settled, agricultural societies who are in a position to accumulate goods. Wealth and poverty, in other words, seem to be inevitable consequences of agriculture and its offspring, civilization.

PART III

GLOBAL HIGHWAYS
OF FOOD

5

SPLINTERS OF PARADISE

We ceased not to buy and sell at the several islands till we came
to the land of Hind, where we bought cloves and ginger and all
manner spices; and thence we fared on to the land of Sind, where
also we bought and sold. In these Indian seas, I saw wonders
without number or count.

—FROM "SINDBAD THE SEAMAN,"
IN *The Book of the Thousand Nights and a Night*,
TRANSLATED BY SIR RICHARD BURTON (1885–88)

THE CURIOUS APPEAL OF SPICES

Flying snakes, giant carnivorous birds, and fierce bat-like creatures
were just some of the perils that awaited anyone who tried to gather
spices in the exotic lands where they grew, according to the historians of ancient Greece. Herodotus, the Greek writer of the fifth
century B.C. known as the "father of history," explained that gathering cassia, a form of cinnamon, involved donning a full-body suit
made from the hides of oxen, covering everything but the eyes.
Only then would the wearer be protected from the "winged creatures like bats, which screech horribly and are very fierce . . . they
have to be kept from attacking the men's eyes while they are cutting
the cassia."

Even stranger, Herodotus claimed, was the process of collecting
cinnamon. "In what country it grows is quite unknown," he wrote.
"The Arabians say that the dry sticks, which we call kinamomon, are
brought to Arabia by large birds, which carry them to their nests,
made of mud, on mountain precipices which no man can climb.

The method invented to get the cinnamon sticks is this. People cut up the bodies of dead oxen into very large joints, and leave them on the ground near the nests. They then scatter, and the birds fly down and carry off the meat to their nests, which are too weak to bear the weight and fall to the ground. The men come and pick up the cinnamon. Acquired in this way, it is exported to other countries."

Theophrastus, a Greek philosopher of the fourth century B.C., had a different story. Cinnamon, he had heard, grew in deep glens, where it was guarded by deadly snakes. The only safe way to collect it was to wear protective gloves and shoes and, having gathered it, to leave one third of the harvest behind as a gift to the sun, which would cause the offering to burst into flames. Yet another tale told of the flying snakes that protected the frankincense-bearing trees. According to Herodotus, the snakes could be driven off by spice harvesters only by smoking them out with burning storax, an aromatic resin, to produce clouds of incense.

Writing in the first century A.D., Pliny the Elder, a Roman writer, rolled his eyes at such stories. "Those old tales," he declared, "were invented by the Arabs to raise the price of their goods." He might have added that the tall stories told about spices also served to obscure their origins from European buyers. Frankincense came from Arabia, but cinnamon did not: Its origins lay much farther afield, in southern India and Sri Lanka, from where it was shipped across the Indian Ocean, along with pepper and other spices. But the Arab traders who then carried these imported products, together with their own local aromatics, across the desert to the Mediterranean in camel caravans preferred to keep the true origins of their unusual wares shrouded in mystery.

It worked brilliantly. The Arab traders' customers around the Mediterranean were prepared to pay extraordinary sums for spices, largely as a result of their exotic connotations and mysterious origins. There is nothing inherently valuable about spices, which are mainly plant extracts derived from dried saps, gums, and resins; barks; roots; seeds; and dried fruits. But they were prized for their unusual scents and tastes, which are in many cases defensive mechanisms to ward off insects or vermin. Moreover, spices are nutritionally superfluous. What they have in common is that they are durable, lightweight, and hard to obtain, and are only found in specific places. These factors made them ideal for long-distance trade—and the farther they were carried, the more sought-after, exotic, and expensive they became.

WHY SPICES WERE SPECIAL

The English word *spice* comes from the Latin *species*, which is also the root of words such as *special, especially,* and so on. The literal meaning of *species* is "type" or "kind"—the word is still used in this sense in biology—but it came to denote valuable items because it was used to refer to the types or kinds of things on which duty was payable. The Alexandria Tariff, a Roman document from the fifth century A.D., is a list of fifty-four such things, under the heading *species pertinentes ad vectigal,* which literally means "the kinds (of things) subject to duty." The list includes cinnamon, cassia, ginger, white pepper, long pepper, cardamom, aloewood, and myrrh, all of which were luxury items that were liable to 25 percent import duty at the Egyptian port of Alexandria, through which spices from the East flowed into the Mediterranean and then on to European customers.

Today we would recognize these kinds of things, or "species," as spices. But the Alexandria Tariff also lists a number of exotic items—lions, leopards, panthers, silk, ivory, tortoiseshell, and Indian eunuchs—that were technically spices, too. Since only rare and expensive luxury items that were subject to extra duty qualified as spices, if the supply of a particular item increased and its price fell, it could be taken off the list. This probably explains why black pepper, the Romans' most heavily used spice, does not appear on the Alexandria Tariff: It had become commonplace by the fifth century as a result of booming imports from India. Today the word *spice* is used in a narrower, more food-specific way. Black pepper is a spice, even though it does not appear on the Tariff, and tigers are not, even though they do.

So spices were, by definition, expensive imported goods. This was a further component of their appeal. The conspicuous consumption of spices was a way to demonstrate one's wealth, power, and generosity. Spices were presented as gifts, bequeathed in wills along with other valuable items, and even used as currency in some cases. In Europe the Greeks seem to have pioneered the culinary use of spices, which were originally used in incense and perfume, and (as with so many other things) the Romans borrowed, extended, and popularized this Greek idea. The cookbook of Apicius, a compilation of 478 Roman recipes, called for generous quantities of foreign spices, including pepper, ginger, putchuk (costus), malabathrum, spikenard, and turmeric, in such recipes as spiced ostrich. By the Middle Ages food was being liberally smothered in spices. In medieval cookbooks spices appear in at least half of all recipes, sometimes three quarters. Meat and fish were served with richly spiced sauces including various combinations of cloves, nutmeg, cinnamon, pepper, and mace. With their richly spiced food, the wealthy literally had expensive tastes.

This enthusiasm for spices is sometimes attributed to their use in masking the taste of rotten meat, given the supposed difficulty of preserving meat for long periods. But using spices in this way would have been a very odd thing to do, given their expense. Anyone who could afford spices could certainly have afforded good meat; the spices were the more expensive ingredient by far. And there are many recorded medieval examples of merchants who were punished for selling bad meat, which rather undermines the notion that meat was invariably putrid and rotten, and suggests that spoiled meat was the exception rather than the rule. The origin of the surprisingly persistent myth about spices and bad meat may lie in the use of spices to conceal the saltiness of meat that had been preserved by the widespread practice of salting.

Spices were certainly regarded as antidotes to earthly squalor in another, more mystical sense. They were thought to be splinters of paradise that had found their way into the ordinary world. Ginger and cinnamon were said by some ancient authorities to be hauled from the Nile in nets, having washed down the river from Paradise (or the Garden of Eden, according to later Christian writers), where exotic plants grew in abundance. They provided an otherworldly taste of paradise amid the sordid reality of earthly existence. Hence the religious use of incense, to provide the scent of the heavenly realm, and the practice of offering spices to the gods as burnt offerings. Spices were also used to embalm the dead and prepare them for the afterlife. The mythical phoenix was even said by one Roman writer to make her nest from—what else?—a selection of spices. "She collects the spices and aromas that the Assyrian gathers, and the rich Arab; those that are harvested by Pygmy peoples and by India, and that grow in the soft bosom of the Sabaean land. She collects cinnamon, the perfume of far-wafting amomum, balsams mixed with tejpat leaves; there is also a slip of gentle cassia and gum arabic, and the rich teardrops of frankincense. She adds the tender spikes of downy nard and the power of Panchaea's myrrh."

The appeal of spices, then, arose from a combination of their mysterious and distant origins, their resulting high prices and value as status symbols, and their mystical and religious connotations—in addition, of course, to their smell and taste. The ancient fascination with spices may seem arbitrary and strange today, but its intensity cannot be underestimated. The pursuit of spices is the third way in which food remade the world, both by helping to illuminate its full extent and geography, and by motivating European explorers to seek direct access to the Indies, in the course of which they established rival trading empires. Examining the spice trade from a European perspective might seem strange, given that Europe occupied only a peripheral position and a minor role in the trade in ancient times. But this served to heighten the mystery and the appeal of spices to

Europeans in particular, ultimately prompting them to uncover the true origins of these strangely appealing dried roots, shriveled berries, desiccated twigs, slivers of bark, and sticky bits of gum—with momentous consequences for the course of human history.

THE SPICE TRADE'S WORLD-WIDE WEB

When a ship was found stranded on the shores of the Red Sea, around 120 B.C., there appeared at first to be no survivors. Everyone on board had starved to death—except, it turned out, for one man, and he was only barely alive. He was given food and water and taken to the Egyptian court in Alexandria where he was presented to King Ptolemy VIII (known as Physcon, or "potbelly," because of his girth). But nobody could understand what the foreign sailor was saying, so the king sent him away to learn some Greek, the official language of Egypt at the time. Not long afterward the sailor returned to the court to tell his story. He explained that he was from India and that his ship had gone off course on its way across the ocean, and had ended up drifting in the Red Sea.

Since the only sea route to India known in Egypt at the time involved hugging the coast of the Arabian peninsula—something Alexandrian sailors were forbidden to do by Arab merchants who wanted to keep the profitable trade with India to themselves—the sailor's reference to a fast, direct route across the open ocean to India was met with disbelief. To prove that he was telling the truth, and no doubt to secure a passage home for himself, the sailor offered to act as the guide for an expedition to India. The king agreed and appointed as its leader one of his trusted advisers, a Greek named Eudoxus who was known for his interest in geography. Eudoxus duly sailed away and returned many months later with a cargo of spices and jewels from India, all of which the king confiscated for himself. Eudoxus later made a second trip to India at the behest of Ptolemy VIII's wife and successor, Cleopatra III. Inspired by the wreckage of what appeared to be a Spanish ship on the east African coast of

Ethiopia, he then became obsessed with the idea that it was possible to sail right around Africa. He sailed along the north coast of Africa and headed into the Atlantic to attempt the circumnavigation, but he was never heard from again.

That, at least, is the story related by Strabo, a Greek philosopher who wrote a treatise on geography in the early first century A.D. Strabo himself was skeptical of the tale: Why did the Indian sailor survive, when his shipmates did not? How did he learn Greek so quickly? Yet the story is plausible, because direct sea trade between the Red Sea and the west coast of India really did open up during the first century B.C., just after the shipwrecked Indian is supposed to have appeared in Alexandria. Until this time only Arab and Indian sailors had known the secret of the seasonal trade winds, which allowed fast, regular passage across the ocean between the Arabian peninsula and the west coast of India. These winds blow from the southwest between June and August to carry ships eastward, and then from the northeast between November and January to carry them westward again. Knowledge of the winds, and Arab control of the overland routes across the Arabian peninsula, gave Indian and Arab merchants a firm grip on the trade between India and the Red Sea. They sold spices and other oriental goods to Alexandrian merchants in markets around the southwestern tip of Arabia. These goods were then shipped up the Red Sea, over land to the Nile, and finally up the Nile to Alexandria itself.

Following in Eudoxus's wake, however, Alexandrian sailors learned how to exploit the trade winds—the details are said to have been worked out by a Greek named Hippalos, after whom the southwesterly wind was named—and were then able to bypass the Arabian markets and sail directly across the ocean to India's west coast, cutting out the Arab and Indian middlemen. The volume of shipping increased as Roman traders gained direct access to the Red Sea following Egypt's annexation by Rome in 30 B.C. Roman control of trade between the Red Sea and India was cemented under the emperor Augustus, who ordered attacks on the ports of southern

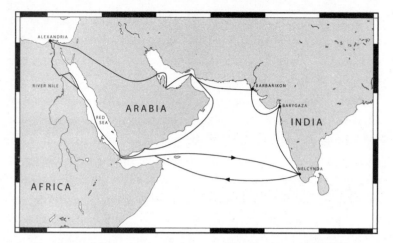

*Knowledge of the sea route to India gave Alexandrian (and later Roman)
sailors direct access to the spice markets of India's west coast,
bypassing Arabia altogether.*

Arabia, reducing Aden, the main market city, to "a mere village" according to one observer. By the early first century A.D. as many as 120 Roman ships a year were sailing to India to buy spices, including black pepper, costus, and nard—along with gems, Chinese silk, and exotic animals for slaughter in the Roman world's many arenas. For the first time Europeans had become direct participants in the thriving trade network of the Indian Ocean, the hub of global commerce at the time.

The "Periplus of the Erythraean Sea," a sailor's handbook written by an unknown Greek navigator in the first century A.D., gives a flavor of the frenetic commercial activity in the markets interconnected by the Indian Ocean. It lists the ports along the west coast of India and their specialties, from Barbarikon in the north (a good place to buy costus, spikenard, bdellium, and lapis lazuli), to Barygaza (good for long pepper, ivory, silk, and a local form of myrrh) and right down to Nelcynda, almost at the southern tip of India. In this region the main trade was in pepper, which was "grown in quantity" inland, according to the Periplus. Also on offer was malabathrum, the leaf of the local cinnamon plant and a particularly valued spice: A pound of small leaves would fetch seventy-five denarii in Rome, or about six times the typical monthly salary. In all these ports Roman traders offered wine, copper, tin, lead, glass, and red coral from the Mediterranean, which was valued in India as a protective charm. But mostly the Roman traders had to pay for spices with gold and silver, since most of their goods had little appeal to Indian merchants. Tamil poems of the first century A.D. refer to the "yavanas," a generic term for people from the west, with their great ships and wealth that "never wane[d]," a reference to the vast quantities of gold and silver that were handed over in return for spices.

The Periplus goes on to tell of the ports on India's east coast and of the small vessels that traded between the east and west coasts. It also mentions the much larger ships that plied the Bay of Bengal between India and southeast Asia, which were probably Malay or Indonesian vessels. Given the size of Roman vessels, the fact that the size of these ships is remarked upon suggests that they were very large indeed. They would have carried goods from farther east, including nutmeg, mace, and cloves from the spice islands of Indonesia (the Moluccas) and silk from China.

Beyond this point the Periplus becomes rather vague. But it provides at least a glimpse, from the European perspective, of a vast trade network, the first connections of which had been established thousands of years earlier. Cardamom from southern India had been available in Mesopotamia in the third millennium B.C., Egyptian ships were bringing frankincense and other aromatics from the land of Punt (probably Ethiopia) in the second millennium B.C., and Pharaoh Ramses II was buried in 1224 B.C. with a peppercorn from India inserted in each of his nostrils. In a wave of expansion between 500 B.C. and 200 A.D., however, the spice-trade network came to encompass the entire Old World, with cinnamon and pepper from India being carried as far west as Britain and frankincense from Arabia traveling as far east as China. But the full extent of this network was generally unknown to its participants, since they were not always aware of the origins of the goods they traded. Just as the

Greeks thought that the Indian spices that reached them via Arab traders actually originated in Arabia, so too the Chinese seem to have assumed that nutmeg and cloves came from Malaya, Sumatra, or Java, though these were in reality just ports of call on the way along the maritime trade routes from their true source farther east, in the Moluccas.

Spices also crossed the world by land. From the second century B.C. overland routes connected China with the eastern Mediterranean, linking the Roman world in the west and Han China in the east. (These routes were dubbed the Silk Road in the nineteenth century, even though they carried far more than silk and there was in fact a network of east-west routes, not a single road.) Musk, rhubarb, and licorice were among the spices traded along this route. Spices also traveled by land between the north and south of India, between India and China, and between southeast Asia and inland China. Nutmeg, mace, and cloves were available in India and China in Roman times but did not regularly reach Europe until the dying days of Roman rule.

The extent of this trade, and the amount spent importing exotic foreign goods, provoked some opposition in Rome. For one thing it was extravagant, which was not in keeping with the supposedly traditional Roman values of modesty and frugality. It also meant that large amounts of silver and gold were flowing east. Compensating for this outflow required that the Romans find new sources of treasure, either through conquest or by opening up new mines. And all of this was for products that were, strictly speaking, unnecessary and were sold at heavily marked-up prices.

As Pliny the Elder put it: "In no year does India absorb less than 55 million sesterces of our wealth, sending back merchandise to be sold to us at one hundred times its prime cost." In total, he reported, Rome's annual trade deficit with the east amounted to one hundred million sesterces, or about ten tons of gold, once Chinese silk and other fine goods were taken into account along with the spices. "Such is the sum that our luxuries and our women cost us," he lamented. Pliny professed to be baffled by the popularity of pepper. "It is remarkable that its use has come into such favor, for with some foods it is their sweetness that is appealing, others have an inviting appearance, but neither the berry nor the fruit of pepper has anything to recommend it," he wrote. "The sole pleasing quality is its pungency—and for the sake of this we go to India!"

Similarly, Pliny's contemporary Tacitus worried about Roman dependence on "spendthrift table luxuries." When he wrote these words around the end of the first century A.D., however, the Roman spice trade was already past its peak. As the Roman Empire declined and its wealth and sphere of influence shrank in the centuries that followed, the direct spice trade with India withered in turn, and Arab, Indian, and Persian traders reasserted themselves as the main suppliers to the Mediterranean. But the spices continued to flow. A Roman cookbook from the fifth century A.D., "The Excerpts of Vinidarius," lists more than fifty herbs, spices, and plant extracts under the heading "Summary of spices which should be in the house in order that nothing is lacking in seasoning," including pepper, ginger, costus, spikenard, cinnamon leaf, and cloves. And when Alaric, king of the Goths, besieged Rome in 408 A.D., he demanded a ransom of 5,000 pounds of gold, 30,000 pieces of silver, 4,000 robes of silk, 3,000 pieces of cloth, and 3,000 pounds of pepper. Evidently the supply of Chinese silk and Indian pepper continued even as the Roman Empire crumbled and fragmented.

But during the period when direct trade with the east had thrived, it briefly brought the people of Europe into the vibrant Indian Ocean trade system. In the first century A.D., this trading network spanned the Old World, linking the mightiest empires in Eurasia at the time: the Roman Empire in Europe, the Parthian Empire in Mesopotamia, the Kushan Empire in northern India, and the Han dynasty in China. (Rome and China even established diplomatic contacts with each other.) Spices were just one of the things that traveled around this global network by land and sea. But since they had a high ratio of value to weight, could only be found in certain

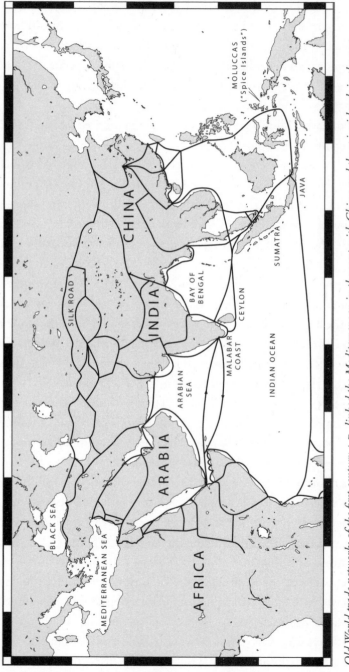

Old World trade networks of the first century A.D. linked the Mediterranean in the west with China and the spice islands in the east.

parts of the world in many cases, were easily stored, and were highly sought after, spices were exceptional in being traded from one end of the network to the other, as shown for example by the references in Roman sources to cloves, which grew only in the Molucca Islands on the other side of the globe. Spices brought a flavor of southeast Asia to Roman tables and the scent of Arabia to Chinese temples. And as spices were traded around the world, they carried other things along with them.

FREIGHTED WITH MEANING

Goods are not the only things that flow along trade routes. New inventions, languages, artistic styles, social customs, and religious beliefs, as well as physical goods, are also carried around the world by traders. So it was that knowledge of wine and wine-making traveled from the Near East to China in the first century A.D.; and knowledge of noodles traveled back in the other direction. Other ideas soon followed, including paper, the magnetic compass, and gunpowder. Arabic numerals actually originated in India, but they were transmitted to Europe by Arab traders, which explains their name. Hellenistic influences are clearly visible in the art and architecture of the Kushan culture of northern India; Venetian buildings were decorated with Arab flourishes. But in two fields in particular—geography and religion—the interplay between trade and the transmission of knowledge was mutually reinforcing.

One of the things that makes spices seem so exotic is their association with mysterious, far-off lands. For early geographers in the ancient world, attempting to put together the first maps and descriptions of the world, spices often marked the boundaries of their knowledge. Strabo, for example, referred to "the Indian cinnamon-producing country" which lay "on the edge of the habitable world," beyond which the earth was, he said, too hot to allow humans to live. Even the more worldly author of the Periplus had little idea what happened east of the mouth of the Ganges: there was a large island, "the last place of the habitable world" (possibly Sumatra), after which "the sea comes to an end somewhere." To the north was the mysterious land of "Thina" (China), the source of silk and malabathrum (cinnamon) leaves.

Traders and geographers depended on each other: Traders needed maps, and mapmakers needed information. Traders would visit geographers before setting out, and might then share information on their return. Knowing how many days it took to travel from one point to another, or typical itineraries of particular routes, made estimates of distance possible, and hence the construction of maps. In this way geographers learned about the layout of the world as an indirect result of the trade in spices and other goods. This is also why so much information about spices comes from the early geographers. Neither they nor the traders wanted to reveal all their secrets, but some give and take made sense for both parties. Merchants worked hand in hand with mapmakers, culminating in the map compiled in the second century A.D. by Ptolemy, a Roman mathematician, astronomer, and geographer. It was surprisingly accurate by modern standards and formed the basis of Western geography for more than a thousand years.

The interdependence between geography and trade was pointed out by Ptolemy himself, who noted that it was only due to commerce that the location of the Stone Tower, a key trading post on the Silk Road to China, was known. He was well aware that the Earth was spherical, something that had been demonstrated by Greek philosophers hundreds of years earlier, and he agonized about how best to represent it on a flat surface. But Ptolemy's estimate of the circumference of the Earth was wrong. Although Eratosthenes, a Greek mathematician, had calculated the circumference of the Earth four hundred years earlier and arrived at almost exactly the right answer, Ptolemy's figure was one-sixth smaller—so he thought the Eurasian landmass extended farther around the world than it actually did. This overestimate of the extent to which Asia extended to the east was one of the factors that later emboldened Christopher Columbus to sail west to find it.

Ptolemy also believed that the Indian Ocean was landlocked, despite reports that it could be reached from the Atlantic by going around the southern tip of Africa. (Herodotus, for example, told of Phoenicians who had circumnavigated Africa around 600 B.C., taking around three years to do so and finding the seasons strangely reversed as they headed south.) Arab geographers realized that the idea of a landlocked Indian Ocean was wrong during the tenth century. One of them, al-Biruni, wrote of "a gap in the mountains along the south coast [of Africa]. One has certain proofs of this communication although no one has been able to confirm it by sight." Al-Biruni's informants were undoubtedly merchants.

Religious beliefs were another kind of information that spread naturally along trade routes, as missionaries followed routes opened up by traders, and traders themselves took their beliefs to new lands. Mahayana Buddhism spread along trade routes from India to China and Japan, and Hinayana Buddhism spread from Sri Lanka to Burma, Thailand, and Vietnam. Tradition has it that Thomas the Apostle took Christianity to India's Malabar coast in the first century A.D., arriving on a spice-trader's ship in Cranganore (modern Kodungallur) in 52 A.D. But trade's most striking religious symbiosis was with Islam. The initial expansion of Islam from its birthplace on the Arabian peninsula was military in nature. Within a century of the death of the prophet Muhammad in 632 A.D., his followers had conquered all of Persia, Mesopotamia, Palestine and Syria, Egypt, the rest of the northern African coast, and most of Spain. But the spread of Islam after 750 A.D. was closely bound up with trade: As Muslim traders traveled outward from the Arab peninsula they took their religion with them.

Arab trading quarters in foreign ports quickly converted to Islam. The African empires that traded with the Muslim world across the Sahara (such as the kingdom of Ghana, and the Mali Empire that replaced it) converted between the tenth and twelfth centuries. Islam also spread along trade routes into the cities of Africa's east coast. And, of course, it was carried along the spice routes of the

Indian Ocean to the west coast of India and beyond. By the eighth century Arab traders were sailing all the way to China to trade in Canton—a direct trade facilitated by political unification brought about by the rise of Islam in the west and the emergence of China's Tang dynasty in the east. But the voyage was a particularly hazardous one. Buzurg ibn Shahriyar, a Persian writer, tells of a captain, Abharah, a legendary navigator who made the voyage to China seven times and lived to tell the tale, but only just: He was shipwrecked on one of his voyages and escaped as the only survivor from his ship.

This is the swashbuckling period depicted in the tales of Sinbad (or Sindbad) the Sailor, of great oceanic voyages, returning home a rich man, spending the spoils, and then becoming restless for adventure and setting out again. Sinbad's tales draw upon the real experiences of Arab traders who plied the Indian Ocean. The direct trade with China ended in 878 A.D., however, when rebels opposed to the Tang regime sacked Canton and killed thousands of foreigners; thereafter merchants from Arabia only went as far as India or southeast Asia, where they traded with Chinese merchants. But Islam continued to spread along the trade routes and eventually took root right around the Indian Ocean, reaching Sumatra in the thirteenth century and the spice islands of the Moluccas in the fifteenth century.

Trade and Islam proved to be highly compatible. Being a merchant was regarded as an honorable profession, not least because Muhammad himself had been one, making several trips to Syria along the overland routes that carried spices from the Indian Ocean into the Mediterranean. As Islam spread, the common language, culture, laws, and customs within the Muslim world provided a fertile environment in which trade could prosper. Visiting Muslim traders were more inclined to do business with coreligionists in trading centers; and once a major trading city in a particular region converted to Islam, it made sense for other towns nearby to follow suit, adopting Muslim laws and the Arabic language. The Venetian explorer Marco Polo, visiting Sumatra in the late thirteenth century, noted that the island's northeast tip was "so much frequented by

Saracen [Arab] merchants that they [had] converted the natives to the Law of Mahomet." Even if some merchants initially converted for reasons of commercial expediency, Islam's rapid spread suggests that they, or at least their descendants, soon became entirely sincere in their embrace of the new religion. Trade spread Islam, and Islam promoted trade. It is worth noting that at the end of the twentieth century, the two countries with the largest Muslim populations were Indonesia and China—both far beyond the realm of Islam's military conquests.

Two historical figures illustrate Islam's reach and unifying power. The first is Ibn Battuta, a Muslim from Tangiers who is often referred to as the Arab Marco Polo. In 1325, at the age of twenty-one, he set out to make the pilgrimage (hajj) to Mecca, where he arrived the following year, having visited Cairo, Damascus, and Medina along the way. But rather than return home directly, he decided to do some more traveling and embarked on what turned into a twenty-nine-year, 73,000-mile journey around much of the known world. He visited Iraq, Persia, the east coast of Africa, Turkey, and Central Asia and traveled across the Indian Ocean to southern China. He then returned to North Africa, from where he visited southern Spain and the central African kingdom of Mali. It was an amazing journey by any standard, but what is particularly remarkable is that for most of his travels, Ibn Battuta remained within the Muslim world, or what Muslims call *dar al-Islam* (literally, "the abode of Islam"). He served as a judge in Delhi and the Maldives, was sent by an Indian sultan as an ambassador to China, and when he visited Sumatra in 1346, he found that the local sultan's jurists were members of his own Hanafi school of legal thought.

The second figure is Zheng He, the admiral of China's extraordinary armada of treasure ships. Between 1405 and 1433 he commanded seven official voyages, each lasting two years, that traveled far into the Indian Ocean. His fleet of 300 ships, manned by 27,000 sailors, was the largest ever assembled, and it was to remain unsurpassed in size for another five hundred years. Zheng He's instructions were to demonstrate China's wealth, might, and sophistication to other nations, establish diplomatic links, and encourage trade. Accordingly, he sailed via the spice islands of southeast Asia to the coast of India, up the Persian Gulf, and as far west as Africa's east coast. Along the way his ships gathered curiosities, traded with local rulers, and collected ambassadors to take back to China. Zheng He was China's ambassador to the outside world; perhaps surprisingly, he was also a Muslim. But that made him ideally qualified to navigate the ports, markets, and palaces of the kingdoms around the Indian Ocean. Ultimately, however, his efforts came to nothing. Although he established China as a powerful presence in the Indian Ocean, internal rivalries within the Chinese court led to the disbanding of the navy, in part to settle political scores, but also so that resources could be diverted instead to protecting the empire from land-based attackers from the north.

If the world's spice-trading networks were the communications networks of their day, linking up far-flung lands, then Islam was the common protocol on which they operated. But although trade flourished in the Muslim world, the rise of Islam had the effect of cutting Europe off from the Indian Ocean trade system. Once Alexandria fell to Muslim troops in 641 A.D., spices could no longer reach the Mediterranean directly: Europeans were relegated to a commercial backwater by a "Muslim curtain" that blocked their access to the east.

AROUND THE MUSLIM CURTAIN

In 1345 Jani Beg, the khan of the Golden Horde, laid siege to the port of Caffa on the Crimean peninsula. Genoese traders had purchased the city from the Golden Horde (the westernmost fragment of the collapsed Mongol Empire) in 1266 and it was their main trading emporium in the Black Sea. But Jani Beg disapproved of the use of the port for slave trading and tried repeatedly to take it back. Just as it looked as though he was about to succeed, however, his army was struck by a terrible plague. According to a contemporary account by Gabriele de Mussi, an Italian notary, Jani Beg's troops loaded plague-ridden corpses into catapults and fired them into the city. The defenders threw the bodies over the walls of Caffa and into the sea, but the plague had taken hold. "Soon, as might be supposed, the air became tainted and the wells of water poisoned, and in this way the disease spread so rapidly in the city that few of the inhabitants had strength sufficient to fly from it," de Mussi recorded. But some of the Genoese did manage to flee—and as they headed westward they took the plague with them in their ships.

The plague, known today as the Black Death, spread throughout the Mediterranean basin during 1347, reaching France and England in 1348 and Scandinavia by 1349, and killing between one third and one half of the population of Europe by 1353, by some estimates. "A plague attacked almost all the sea coasts of the world and killed most of the people," noted a Byzantine chronicler. The exact biological nature of the plague is still hotly debated, though it is generally thought to have been bubonic plague, carried by fleas on black rats. It was known at the time as the "pestilence"; the term "Black Death" was coined in the sixteenth century and became popular in the nineteenth. No treatment could save victims once the plague took hold. There are accounts of people being sealed into their houses to prevent the plague from spreading, and of people abandoning their families to avoid infection. Medical men proposed all sorts of strange measures that would, they said, minimize the risk of infection, advising fat people not to sit in the sunshine, for example, and issuing a baffling series of dietary pronouncements. Doctors in Paris advised people to avoid vegetables, whether pickled or fresh; to avoid fruit, unless consumed with wine; and to refrain from eating poultry, duck, and meat from young pigs. "Olive oil," they warned, "is fatal."

Among the long lists of foods to avoid, there were a few examples of foods that were meant to offer protection from the plague—chief among them spices, with their exotic, quasi-magical associations, pungent aromas, and long history of medical uses. The French doctors recommended drinking broth seasoned with pepper, ginger, and cloves. The plague was thought to be caused by corrupted air, so people were advised to burn scented woods and sprinkle rosewater in their homes, and to carry various concoctions of pepper, rose petals, and other aromatics when going out. The Italian writer Giovanni Boccaccio described people who "walked abroad, carrying in their hands flowers or fragrant herbs or divers sorts of spices, which they frequently raised to their noses." This helped to conceal the smell of the dead and dying, as well as supposedly purifying the air. John of Escenden, a fellow at Oxford University, was certain that a combination of powdered cinnamon, aloes, myrrh, saffron, mace, and cloves had enabled him to survive even as those around him succumbed to the plague.

But as a means of preventing infection spices were, in fact, completely useless. Indeed, they were worse than useless; they were partly to blame for the arrival and spread of the plague in the first place. The Genoese port of Caffa was valuable because it sat at the western terminus of the Silk Road to China, and because spices and other goods from India, shipped up the Gulf and then carried overland to Caffa and other Black Sea ports, went around the back of the Muslim curtain. So Caffa allowed the Genoese to circumvent the Muslim monopoly and obtain eastern goods for sale to European customers. (Their arch-rivals, the Venetians, had by this time allied themselves with the Muslim sultans who controlled the Red Sea trade, and acted as their official European distributors.) The plague, which appears to have originated in central Asia, reached Caffa along the overland trade routes before being spread around Europe by Genoese spice ships.

By the time the connection between the spice trade and the plague was noticed, it was too late. "In January of 1348 three galleys put in at Genoa, driven by a fierce wind from the East, horribly infected and laden with a variety of spices and other valuable goods," wrote a Flemish chronicler. "When the inhabitants of Genoa learnt this, and saw how suddenly and irremediably they infected other people, they were driven forth from that port by burning arrows and divers engines of war; for no man dared to touch them; nor was any man able to trade with them, for if he did he would be sure to die forthwith. Thus they were scattered from port to port." Later that year a French writer in Avignon wrote of the Genoese ships that "people do not eat, nor even touch spices, which have not been kept a year, since they fear they may lately have arrived in the aforesaid ships . . . it has many times been observed that those who have eaten the new spices . . . have suddenly been taken ill."

The relative importance of the various land and sea routes between Europe and the East varied in accordance with the geopolitical situation in central Asia. Political unification under the Mongol Empire, for example, which encompassed much of the northern Eurasian landmass, from Hungary in the west to Korea in the east, made overland trade much safer, and volumes increased accordingly: In the thirteenth century it was said that a maiden could walk across the Mongol Empire with a pot of gold on her head without being molested. The establishment of Christian toeholds in the Levant during the Crusades provided other outlets for goods brought overland along the Silk Road or from the Gulf. Conversely, the breakup of the Mongol Empire in the early fourteenth century meant that the balance tipped back in favor of the Red Sea route, now controlled by the Muslim dynasty of the Mamluks.

During the fifteenth century there was increasing concern in Europe over the extent of Muslim control over trade with the east. By 1400 some 80 percent of this trade was in Muslim hands. Their European distributors, the Venetians, were at the height of their powers. Venice handled around five hundred tons of spices a year, around 60 percent of which was pepper. The cargo of a single Venetian galley was worth a royal ransom. Various popes tried to ban trade with the Muslim world, but the Venetians either ignored them or won special dispensations to continue doing business as usual. Genoa, meanwhile, was in decline. Its Black Sea possessions were under pressure from the Ottoman Turks, a rising Muslim power that was encroaching upon the fast-shrinking Byzantine Empire. And between 1410 and 1414 there was a sudden spike in the price of spices—in England, the price of pepper increased eightfold—which painfully reminded everyone just how dependent they were on their suppliers. (The cause of this spike was probably the activities of Zheng He, whose unexpected arrival on the west coast of India disrupted the usual patterns of supply and demand and drove up prices.) All of this fueled a growing interest in the possibility of finding some new way around the Muslim curtain and establishing direct trading links with the East.

The fall of Constantinople in 1453 is sometimes portrayed as the event that ultimately triggered the European age of exploration, but it was merely the most prominent in a series of events that finally choked off the land route to the East altogether. The Ottoman Turks had already conquered Greece and most of western Turkey by 1451, and they regarded Constantinople, by now the last significant holdout of the old Byzantine Empire, as "a bone in the throat of Allah." Once it had fallen they imposed huge tolls on ships entering and leaving the Black Sea, and then went on to take the Genoese ports around its coast, including Caffa, which fell in 1475. Meanwhile the Ottomans' Muslim rivals, the Mamluks, took the opportunity to raise the tariffs on spices passing through Alexandria, causing prices in Europe to increase steadily during the second half of the fifteenth century. It was not simply the fall of a single city, in short, but the slow crescendo of concern over the Muslim spice monopoly that prompted European explorers to seek radical new sea routes to the East.

6

SEEDS OF EMPIRE

After the year 1500 there was no pepper to be had at Calicut
that was not dyed red with blood.

—VOLTAIRE, 1756

"I BELIEVE I HAVE FOUND RHUBARB AND CINNAMON"

In June 1474 Paolo Toscanelli, an eminent Italian astronomer and
cosmographer, wrote a letter to the Portuguese court in Lisbon out-
lining his unusual theory: that the fastest route from Europe to India, "the land of spices," was to sail west, rather than trying to sail
south and east around the bottom of Africa. "And be not amazed
when I say that spices grow in lands to the west, even though we
usually say the east," he wrote. Toscanelli described the riches of the
east, borrowing heavily from Marco Polo's account, and helpfully
included a nautical chart showing the islands of Cipangu and Antil-
lia in the ocean on the way to Cathay (China), which he estimated
to be 6,500 miles to the west of Europe. "This country is richer than
any other yet discovered, and not only could it provide great profit
and many valuable things, but it also possesses gold and silver and
precious stones and all kinds of spices in large quantities," he declared. The Portuguese court ultimately ignored Toscanelli's advice,
but Christopher Columbus, a Genoese sailor living in Lisbon at the
time, heard of his letter and obtained a copy of it, possibly from
Toscanelli himself.

Columbus, like Toscanelli, was convinced that sailing west was the
fastest route to the Indies, and he spent years amassing documents

that supported his case, performing calculations, and drawing maps. The idea had solid intellectual foundations—the ancient authorities Ptolemy and Strabo had alluded to it—and Columbus also drew inspiration from Pierre d'Ailly, a fourteenth-century French scholar whose "Description of the World" declared that the journey from Spain to India, sailing west, would take "a few days." But the backing of Toscanelli, one of the most respected cosmographers of his day, gave the theory added weight.

Building on the calculations of Ptolemy, who had overestimated the size of Eurasia and underestimated the circumference of the Earth, Columbus cherry-picked figures from various authorities to convince himself that the Earth was even smaller and Eurasia even bigger, thus shrinking the intervening ocean. He used an estimate from al-Farghani, a Muslim geographer, for the circumference of the Earth; but he failed to appreciate the difference between Muslim and Roman miles and ended up with a figure that was, conveniently, 25 percent too small. Then he used Marinus of Tyre's unusually large estimate of the size of Eurasia, and added on Marco Polo's reports of Cipangu (Japan), a large island said to be hundreds of miles off the east coast of China, which further reduced the width of the ocean he would have to cross. In this way Columbus calculated the distance from the Canary Islands (off Africa's west coast) to Japan to be slightly over two thousand miles—less than a quarter of the true figure.

Convincing a patron to back his proposed expedition proved difficult, however. This was not, as is sometimes suggested, because the panels of experts appointed in the 1480s by the Portuguese and Spanish courts to evaluate Columbus's proposal disagreed with his contention that the Earth was spherical; that was generally accepted. The problem was that his calculations looked fishy, particularly since they relied on evidence from Marco Polo, whose book describing his travels in the East was widely regarded at the time as a work of fiction. Portugal was, in any case, pursuing its own program of exploration down the west coast of Africa, and was unwilling to abandon it (which is why Toscanelli's letter also fell on deaf ears). So both panels of experts said no. But Columbus's fortunes changed when King Ferdinand and Queen Isabella of Spain, fresh from their victory at Granada, the last Muslim stronghold in Spain, decided to back him after all. Columbus may have swayed them by suggesting that the proceeds of his expedition could fund a campaign to reconquer Jerusalem. He certainly presented his voyage as an unashamedly commercial venture, and the documents defining the terms of the expedition granted him "a tenth of all gold, silver, pearls, gems, spices and other merchandise produced or obtained by barter and mining within the limits of these domains."

His three ships headed west from the Canary Islands on September 6, 1492, and encountered land, after an increasingly anxious voyage, on October 12. Columbus was certain that riches were in his grasp as soon as land was sighted. His log refers repeatedly to "gold and spices" and details his attempts to get the local people to tell him where to find them. "I was attentive and took trouble to ascertain if there was gold," he wrote in his log on October 13, after meeting a group of natives. Two weeks after arriving, having visited several among what he took to be the 7,459 islands that Marco Polo claimed lay off the eastern coast of China, he wrote in his log: "I desired to set out today for the island of Cuba . . . my belief being that it will be rich in spices." Columbus failed to find spices on Cuba, but he was told that cinnamon and gold could be found to the southeast. By mid-November he was still maintaining in his log that "without doubt there is in these lands a very great quantity of gold . . . stones, precious pearls and infinite spicery." In December, lying off the island he had named Hispaniola, he recorded that he could see on the shore "a field of trees of a thousand kinds, all laden with fruit . . . believed to be spices and nutmegs."

Given that Columbus communicated with the local people using sign language, he could interpret their signs in almost any way he chose. Just as conveniently, there were several plausible explanations for his failure to find any spices. Perhaps it was the wrong season; his men did not know the correct harvesting and processing techniques; and of course Europeans did not know what spices looked like in the wild anyway. "That I have no knowledge of the products causes me the greatest sorrow in the world, for I see a thousand kinds of trees, each one with its own special trait, as well as a thousand kinds of herbs with their flowers; yet I know none of them," wrote Columbus. He also suffered from bad luck, it seemed: One crew member said he had found mastic trees, but unfortunately he had dropped the sample; another said he had discovered rhubarb but could not harvest it without a shovel.

Columbus departed for Spain on January 4, 1493, having amassed a small amount of gold through trading with the local people. He also carried back samples of what he took to be spices. After a difficult voyage he arrived back in Spain in March 1493, and his official letter to Ferdinand and Isabella, reporting his discoveries, became a bestseller across Europe, with eleven editions published by the end of that year. He described exotic islands with lofty mountains, strange birds, and new kinds of fruit. On the island of Hispaniola, he wrote, "there are many spiceries, and great mines of gold and other metals." He explained that delivery of the riches of these new lands could start right away: "I shall give their highnesses spices and cotton at once, as much as they shall order to be shipped, and as much as they shall order to be shipped of mastic . . . and aloes as much as they shall order to be shipped; and slaves as many as they shall order to be shipped, and these shall be from idolatrous peoples. And I believe I have found rhubarb and cinnamon."

Judging by the triumphant tone of his letter, it seemed that Columbus had achieved his objective of finding a new route to the riches of the east. Although the islands he visited did not match the descriptions of China and Cipangu from Marco Polo's account, he was confident the mainland was nearby. What better proof than the presence of cinnamon and rhubarb, which were known to originate in the Indies? But opinion in the Spanish court was divided. The twigs that Columbus claimed were cinnamon did not smell right and seemed to have gone bad in the course of the return voyage. His other samples of spices were similarly unimpressive, and he had only found a small quantity of gold. Skeptics concluded that he had found nothing more important than a few new Atlantic islands. But Columbus claimed to be closing in on the source of the gold, so a second, much larger expedition was dispatched.

The second expedition only perpetuated the confusion over the presence of spices. Writing home to Seville from Hispaniola in 1494, Diego Álvarez Chanca, who acted as Columbus's doctor on the voyage, explained the situation. "There are some trees which 'I think' bear nutmegs but are not in fruit at present. I say 'I think' because the smell and taste of the bark resembles nutmegs," he wrote. "I saw a root of ginger, which an Indian had tied round his neck. There are also aloes: it is not of a kind which has hitherto been seen in our country, but I am in no doubt that it has medicinal value. There is also very good mastic." Not one of these things was really there; but the Spanish really wanted them to be. "There is also found a kind of cinnamon; it is true that it is not so fine as that which is known at home," wrote Chanca. "We do not know whether by chance this is due to lack of knowledge of when it should be gathered, or whether by chance the land does not produce better."

Columbus threw himself into exploration, hoping to show that he had found the Asian mainland. He claimed to have found the footprints of griffins and thought he detected similarities between local place names and those mentioned by Marco Polo. At one point he got every sailor in his fleet to swear an oath that Cuba was bigger than any known island, and that they were very close to China. Any sailor who refuted these claims was threatened with a large fine and the loss of his tongue. But doubts grew as Columbus returned from each of his voyages with a few lumps of gold and more of his dubious spices. He fell back on religious justifications for his activities—the natives could be converted to Christianity—though he also suggested that they might make good slaves. His settlers became increasingly rebellious. Columbus was accused of mismanagement of his colonies, and of having painted a misleading picture of their potential. At the end of the third voyage he was sent back to Spain in chains and was stripped of his title as governor. After a fourth and final voyage, he died in 1506, convinced to the end that he had indeed reached Asia.

The idea of finding spices in the Americas outlived Columbus. In 1518 Bartolomé de las Casas, a Spanish missionary to the New World, claimed that the new Spanish colonies were "very good" for ginger, cloves, and pepper. The conquistador Hernán Cortés found lots of gold, plundering it from the Aztecs in the course of the Spanish conquest of Mexico, but even he felt bad about his failure to deliver any nutmeg or cloves. He insisted in letters back to the king of Spain that he would, in time, find the route to the spice islands. In the 1540s another conquistador, Gonzalo Pizarro, scoured the Amazon jungle in a doomed search for the legendary city of El Dorado and the "*país de la canela*," or cinnamon country. It was not until the seventeenth century that the search for Old World spices in the Americas was finally abandoned.

Of course, the Americas offered the rest of the world all kinds of new foodstuffs, including maize, potatoes, squash, chocolate, tomatoes, pineapples, and new flavorings, including vanilla and allspice. And though Columbus failed to find the spices he sought in the New World, he found something that was, in some respects, even better. "There is plenty of aji," he wrote in his log, "which is their pepper, which is more valuable than black pepper, and all the people eat nothing else, it being very wholesome. Fifty caravels might be annually loaded with it." This was the chile, and although it was not pepper, it could be used in a similar way. An Italian observer at the Spanish court noted that five grains were hotter and had more flavor than twenty grains of ordinary pepper from Malabar. Better still, the chile could be grown easily outside its region of origin, unlike most spices, so it quickly spread around the world and had been assimilated into Asian cooking within a few decades.

But despite the chile's culinary virtues, it was not what Columbus wanted. The ease with which it could be transplanted from one region to another meant it did not have the financial value of traditional spices, which was due in large part to the geographical limitations of their supply and the need for long-distance transport. More importantly, however, Columbus wanted to find the Old World spices not simply for their taste or value, but because he wanted to prove that he really had arrived in Asia. That was why he sowed confusion for centuries to come by calling chiles "peppers" and the people he found in the Bahamas "Indians," in each case naming them after what he had set out to find. For to find the source of spices was to have arrived in the Indies, the exotic and aromatic lands described by Marco Polo and others whose tales had bewitched Europeans for so many centuries.

"Christians and Spices"

Spices were not one of the original goals of the Portuguese program to explore the west coast of Africa, which was launched in the 1420s by Infante Henrique of Portugal (known in English as Prince Henry the Navigator, yet another nineteenth-century coinage). Henry's aims were to learn more of the geography of the coast and nearby islands, establish trade links, and perhaps make contact with Prester John, the legendary Christian ruler of a kingdom thought to be somewhere in Africa or the Indies, who would be a valuable ally against the Muslims. As Henry's ships worked their way down the African coast, each going a little farther than the last, they disproved the ancient Greek notion that the earth eventually became too hot for human habitation. They brought back gold, slaves, and "grains of paradise," an inferior pepper-like spice that was vaguely known in Europe since it was sometimes traded across the Sahara to the Mediterranean. They looked for an outlet of the Nile, in the hope of following it upstream to find Prester John. But as the fifteenth century progressed, the European need to find an alternative route to the Indies became steadily more urgent. The Portuguese ships pushed south and eventually, in 1488, Bartholomeu Dias rounded Africa's southern cape by accident after being swept out into the Atlantic by a storm and then heading east. He returned to Lisbon with the news that contrary to the opinion of some of the ancients, the Indian Ocean was not landlocked and could be reached from the Atlantic—and so, by extension, could India.

So why did it take nearly nine years for Portugal to send an expedition to India? Organizing a fleet would have taken time, but Columbus's discoveries in the Atlantic may also have been responsible for the delay. If he really had found a westerly route to the east, then going all the way around Africa would be unnecessary. But when Columbus returned from his second voyage in 1496 with very little to show for it, the Portuguese regained their enthusiasm for an expedition to India around the southern tip of Africa. The ships sailed the following year. As a chronicler of the time succinctly put it: "In the year 1497, the King Manuel, the first of that name in Portugal, sent four ships out, which left on a quest for spices, captained by Vasco da Gama."

The voyage was characterized by religious confusion and rivalry. Having rounded the cape and worked their way up Africa's east coast, da Gama and his men were mistaken for Muslims by the sultan of Mocobiquy (Mozambique). He promised to provide them with a pilot who could guide them to India, but then realized his error. A fight ensued and da Gama's ships bombarded the town, killing at least two people. Further run-ins with local Muslims followed as the Portuguese tried in vain to get hold of a pilot. At Malindi, farther up the African coast, da Gama then mistook the Hindu residents for Christians of an unknown sect. After picking up an expert pilot the Portuguese ships then headed across the Indian Ocean to India's Malabar coast, where they anchored near Calicut (modern Kozhikode) on May 20, 1498. As was customary da Gama sent ashore a *degredado*, usually a criminal or an outcast who was deemed to be expendable, to make contact with the locals. The Indians could not understand him and took him to the house of some resident Muslim merchants from Tunisia. "What the devil brought you here?" they asked the man. "We came in search of Christians and spices," he replied.

Though the latter were clearly present in abundance in Calicut, the former were not. But da Gama and his men thought otherwise, assuming that the local Hindus were Christians, falling to their knees in Hindu temples, and mistaking depictions of Hindu goddesses for the Virgin Mary and images of Hindu gods for Christian saints. The king, or *zamorin*, of Calicut was assumed to be a Christian too, and therefore a natural ally against the resident Muslim traders. But he was deeply unimpressed by the trinkets the Portuguese offered (red hats and copper vessels, which were standard trade items on the west coast of Africa). He may have had a distant memory of the appearance in Calicut of Zheng He's treasure fleets, just a few decades earlier, which had offered rich silks in return for spices; more was expected of mysterious foreigners than da Gama's paltry offerings. Da Gama attributed the zamorin's disappointment to the malign influence of the Muslims, and claimed that his ships were merely the vanguard of a much larger treasure fleet, which of course never materialized. So he headed home with only small amounts of pepper, cinnamon, cloves, and ginger, arriving back in Lisbon in September 1499. Only two of his ships and fewer than half his men had survived the voyage—but da Gama's expedition had shown that it was possible to circumvent the Muslim curtain and obtain spices directly from India.

King Manuel was delighted, and was soon styling himself "Lord of Guinea and of the Conquest, the Navigation and the Commerce of Ethiopia, Arabia, Persia, and India." This was of course an enormous exaggeration, but it left no doubt of his intent: to wrest control of the spice trade from the Muslims. Manuel spelled this out in a gloating letter to Ferdinand and Isabella of Spain, in which he explained that his explorers "did reach and discover India and other kingdoms bordering upon it . . . entered and navigated its sea, finding large cities . . . and great populations among whom is carried on all the trade in spices and precious stones." He went on to express his hope that "with the help of God the great trade which now enriches the Moors of these parts . . . shall in consequence of our own regulations be diverted to the natives and ships of our own kingdom so that henceforth all Christendom in this part of the world shall be provided with these spices." Manuel wanted to establish a Portuguese spice monopoly, in short, ostensibly for religious reasons—though obviously there would be commercial benefits too.

Yet how could tiny Portugal hope to displace the throng of Muslim ships in the Indian Ocean, thousands of miles away? Da Gama's men had counted "about fifteen hundred Moorish vessels arriving in search of spices" during the three months they spent in Calicut. But they had also noticed something rather interesting about these ships: they were unarmed. This was standard practice in the Indian Ocean, where there was no dominant political or military power; even the Muslims were divided into several distinct communities. Instead what united the region was trade, based around a handful of major ports and a few dozen smaller ones. In each port traders from different communities could find warehouses to store their goods, banking services, access to local markets, and often a quarter of the city where their fellows resided and their own laws applied. Ports competed to offer the lowest tariffs and attract the highest volume of trade. There was a strong sense of reciprocity: If the police in a particular port mistreated foreign merchants, local merchants were just as likely to complain, since such behavior might lead to retaliation in other ports and undermine trade, which would be bad for everyone. Occasionally local rulers might try to control trade within a particular area using force; but all that did was divert business elsewhere. So unarmed trade was the norm.

Portugal could have gone along with this system, paying Asian authorities for the use of port facilities and handing over tariffs in the usual way. But the Portuguese were used to the way things worked in the Mediterranean, where the use of force to protect sea lanes, shipping, and trading colonies had prevailed since Greco-Roman times. Besides, Portugal did not merely hope to participate in the Indian Ocean trade; it wanted to dominate it, and force the Muslims out. All this soon became apparent during the second Portuguese voyage to India, consisting of thirteen ships under the command of Pedro Alvarez Cabral, which set out in March 1500, less than six months after da Gama's return. As the ships headed south and west into the Atlantic they made an unexpected landfall on the thitherto unknown South American mainland, thus claiming Brazil for Portugal—another unexpected consequence of the search for spices. One ship went back to Lisbon with the news, while the rest pressed on around the African coast, arriving in Calicut in September. Hostilities began almost at once: Cabral's men captured some Muslim ships, and in response the Muslims seized and killed around forty Portuguese merchants in the town. Cabral responded by seizing more Muslim ships and setting them on fire with their crews still aboard. Next, his ships bombarded Calicut for two days, terrifying the inhabitants, before moving on to the ports of Cochin (modern Kochi) and Cannanore (modern Kannur) where the local rulers, keen to avoid a similar fate, allowed the Portuguese to establish trading posts on generous terms.

Cabral's ships then headed back to Portugal, laden with spices. His arrival in July 1501 was greeted with jubilation in Lisbon and dismay in Venice. "This was considered very bad news for Venice," noted one chronicler. "Truly the Venetian merchants are in a bad way." For as well as bringing the first big shipment of spices around the Muslim curtain to Europe, the Portuguese also seemed to have disrupted the Red Sea supply. In 1502 Venetian ships arriving in the Mamluk ports of Beirut and Alexandria found that there was very little pepper to be had, causing prices to rocket and prompting some observers to forecast the ruin of Venice. The number of galleys in its merchant fleet was reduced from thirteen to three, and rather than sending its fleet to Alexandria twice a year, as had previously been the custom, Venice started sending the fleet every other year instead.

Portuguese belligerence reached new heights in the course of the third voyage to India, commanded by Vasco da Gama. His ships ransacked ports on Africa's east coast, exacting booty and demanding tribute. On arrival in India, da Gama arbitrarily burned and bombarded towns on the coast in order to force key ports to buy a *cartaz* from him. This was a permit that granted protection to the port and its ships, and it was only issued on payment of a fee and with a promise not to trade with Muslims—a protection racket, in other words. Da Gama and his men also sank and looted Muslim and local vessels, on one occasion using prisoners for crossbow practice; the hands, noses, and ears of the remaining prisoners were cut off and sent ashore in a boat, and the mutilated people were tied up and burned to death in one of their own ships. Finally, Da Gama negotiated an agreement with the pepper suppliers in Cochin, loaded up with spices, and headed home, sinking a local fleet that had been sent to exact revenge and bombarding Calicut once again for good measure on the way.

This set the tone for the Portuguese efforts to control Indian Ocean trade; any ship or port without a cartaz was deemed to be fair game, local rulers were intimidated into trading on terms generous to the Portuguese, and violence was used arbitrarily and unsparingly. Further expeditions were sent by King Manuel with orders to establish bases in key locations and harass Muslim ships traveling between India and the Red Sea, so that "they are not able to carry any spices to the territory of the [Mamluk] sultan and everyone in India would lose the illusion of being able to trade with anyone but us." Portugal took Goa, on India's west coast, in 1510, making it its main base in the Indian Ocean, and the following year took Malacca, the main distribution point for nutmeg and cloves from the mysterious spice islands, the Moluccas, which lay farther east. Soon afterward a Portuguese expedition finally reached those islands, which had been sought for so long, and informal trade relations were established. Nutmegs and mace were to be found on the nearby Banda islands.

The Portuguese had found the very sources of the spice trade, but their plan to take control of Europe's spice supply from the Muslims ultimately failed. The Indian Ocean was simply too big. At best, Portugal controlled some 10 percent of the Malabar pepper trade and perhaps 75 percent of the flow of spices to Europe, but its attempts to blockade Muslim shipping were never more than partially effective, and by 1560 the flow of spices taken by Muslim traders to Alexandria had recovered to their previous levels. But even though Portugal failed in its efforts to establish a spice monopoly, it did succeed in defining a new model for European trade in the East, based on monopolies and blockades enforced by armed ships from a network of trading posts, which was quickly adopted by its European rivals. Appropriately enough, the rivalries between these emerging colonial powers centered on the Moluccas themselves.

THE SEEDS OF EMPIRE

Spices helped to lure Columbus westward, where none were to be found, and da Gama eastward, where they could be found in abundance. And as if to crown their achievements in establishing new sea routes, spices also inspired the first circumnavigation of the earth. In 1494 Spain and Portugal signed the Treaty of Tordesillas, which included a simple way to divide up the new lands reached by their explorers. They ruled a line down the middle of the Atlantic Ocean, halfway between the Cape Verde islands off the African coast (which were claimed by Portugal) and Hispaniola (which Columbus had just claimed for Spain). Any new lands to the west of the line, it was agreed, would belong to Spain, and those to the east would belong to Portugal; the opinions of the inhabitants were considered to be irrelevant. It subsequently transpired that part of South America, unknown at the time of the treaty, lay to the east of the line, but the agreement clearly stated that it belonged to Portugal, so Portuguese it became. It all seemed very neat and tidy until the Portuguese reached the Moluccas, on the other side of the world. Which side of the line were they on? The 1494 treaty had not specified a dividing line in the Pacific, but the logical way to draw one was to extend the Atlantic meridian right around the earth—in which case, Spain suspected, the spice islands might fall into the hemisphere it considered its property. A Spanish expedition was duly dispatched to establish the precise location of the spice islands and claim them for the Spanish crown.

Oddly enough the expedition was led by a Portuguese navigator, Ferdinand Magellan, who had fallen from favor in the Portuguese court, renounced his nationality, and offered his services to Spain instead. His ships headed west across the Atlantic in 1519 and were the first to cross from the Atlantic to the Pacific via the passage now known as the Strait of Magellan, at the southern tip of South America. Magellan himself was killed in the Philippines in 1521 when he intervened in a dispute between two local chieftains, but the expedition sailed on and reached the Moluccas.

After loading up with cloves, one of Magellan's ships, the *Victoria*, captained by Juan Sebastian Elcano, then continued westward to arrive back in Seville in 1522. The 26 tons of cloves on board covered the entire cost of the expedition, and Elcano was awarded a coat of arms embellished with cinnamon sticks, nutmegs, and cloves. The voyage had proved decisively that the world was round and that the oceans were connected. A crew member on the voyage, a wealthy Italian named Antonio Pigafetta, who kept a detailed diary, also noted something unusual when the ship stopped for supplies at the Cape Verde islands on the way back to Spain: It was the wrong day, "for we had always made our voyage westward and had returned to the same place of departure as the sun, wherefore the long voyage had brought the gain of twenty-four hours, as is clearly seen." But the circumnavigation did not resolve the dispute over the ownership of the Moluccas. That was eventually decided by another treaty, in 1529, when the Spanish abandoned their geographically dubious claim in return for a payment of 350,000 gold ducats from Portugal. And ultimately the question of who was entitled to the Moluccas was rendered moot by the union of the crowns of Spain and Portugal in 1580.

By this time, however, the English and the Dutch had appeared on the scene. The English explorer Francis Drake passed through the Moluccas in 1579 and observed that they yielded an "abundance of cloves, whereof wee furnished our selves of as much as we desired at a very cheap rate." Drake's voyage inspired several follow-up attempts by other English sailors, though all ended in failure. The Dutch were more successful. For a while Dutch merchants had been the distributors for Portuguese spices in northern Europe, but they lost this privilege following Spain's union with Portugal, so they set out to establish their own supply. Intelligence gathered by Jan Huyghen van Linschoten, a Dutch expert on the Indies who had worked for the Portuguese in India for many years, indicated that excellent local pepper was available on Java; and since the Portuguese did not trade there, but bought their pepper in India, they could hardly complain if the Dutch expressed an interest in it. After a successful expedition to Java in 1595, Dutch merchants, who were amalgamated to form the Vereenigde Oost-Indische Compagnie (VOC) or Dutch East India Company in 1602, began regular shipments of spices from the region, exploiting Portugal's inability to control the supply.

Once they realized how tenuous the Portuguese grip really was, the commercially savvy Dutch decided to try to seize control of the trade themselves, and they sent a large fleet to the spice islands in 1605. "The Islands of Banda and the Moluccas are our main target," the VOC's directors explained to their admiral in the region. "We recommend most strongly that you tie these islands to the Company, if not by treaty then by force!" The Dutch ejected the Spanish and Portuguese from the Moluccas, ordered some newly arrived English ships to leave, and seized direct control of the clove supply. The VOC then set about ruthlessly enforcing its new monopoly, determined to succeed where the Portuguese had failed. Clove production was concentrated on the central islands of Ambon and Ceram so that it could be more tightly controlled; the ancient groves of clove trees on other islands were uprooted, the clove pickers massacred, and their villages burned down.

Where clove production was permitted, the growing of other crops was outlawed, to ensure that the local people would be dependent on the Dutch for their food. The Dutch sold the food at a high price and bought the cloves at a low price; even so, production of cloves declined, prompting the Dutch to order that more trees be planted. But by the time the trees came to maturity, supply out-stripped demand, and the growers were told to cut trees down again. A boom-bust cycle followed as the Dutch struggled to reconcile shifting demand with the supply from slow-growing trees and reluc-tant growers. Cultivation of cloves outside Dutch control was for-bidden on pain of death, and clandestine trading was suppressed. Makassar, a regional trading center where the English, Portuguese, and Chinese went to buy smuggled cloves, was shut down.

It was a similar story in the Banda islands, the nearby source of nutmeg and mace. Initially the Dutch persuaded the inhabitants to sign documents agreeing not to sell their spices to anyone else. But they continued to do so anyway, perhaps because they were unaware of what they had signed. In particular, they sold to the English, who had established a base on the tiny island of Run, a little way to the west. A Dutch attempt to build a fort in the Bandas in 1609 pro-voked a dispute with the locals, and a party led by a Dutch admiral who went to negotiate was wiped out by the Bandanese, with the en-couragement of the English. The Dutch retaliated by seizing the Bandas for themselves, building two forts and claiming another spice monopoly. Villages were burned down and the inhabitants were killed, chased off, or sold into slavery. The village chiefs were tortured and then beheaded by the VOC's samurai mercenaries, brought in from Japan, where the Dutch were the only Europeans allowed to trade. The islands were then divided into sixty-eight plots, which were manned with slaves and leased to former VOC employees. The conditions were brutal—workers on the nutmeg plots were executed in a variety of gruesome ways for the most mi-nor transgressions—but the flow of the most valuable spices was now in Dutch hands.

The English agreed to leave the spice islands in 1624 and concen-trated on commercial opportunities in China and India instead, though the Dutch allowed them to retain sovereignty over Run, where a small contingent had held out for many years. This tiny speck of land, two miles long and less than a half mile wide, had originally been claimed by the English in 1603, just as the English and Scottish thrones were united—so it was the first British colonial possession anywhere in the world, and the first tiny step toward the formation of the British Empire. Eventually, in 1667, Run was relinquished to the Dutch under the terms of a Treaty of Breda, one of many peace treaties signed during the on-off Anglo-Dutch wars of the seventeenth and eighteenth centuries. As part of the 1667 deal, Britain received a small island in North America called Manhattan.

Profits from the spice trade helped to bankroll the Dutch "golden age" of the seventeenth century, a period in which the Dutch led the world in commerce, science, and financial innovation, and the wealthy merchant class provided sponsorship for artists such as Rembrandt van Rijn and Johannes Vermeer. But in the long run the Dutch spice monopoly proved to be less valuable than expected. The garrisons and warships needed to protect the monopoly were hugely expensive and did not justify the returns as the price of spices began to fall in Europe in the late seventeenth century. The falling price was due in part to a more abundant supply, so the Dutch imposed artificial constraints on it: They burned huge quantities of spices on the docks in Amsterdam and began to limit the volumes shipped from Asia in an effort to prop up prices. But as trade in textiles became more important, spices accounted for a shrinking proportion of Dutch returns, falling from 75 percent in 1620 to 23 percent in 1700.

The lower prices commanded by spices in Europe also reflected a deeper shift in the spice trade. Once the myths about their otherworldly provenance had been dispelled, spices no longer seemed so glamorous; they started to become affordable, even mundane. Heavily spiced dishes came to be seen as old-fashioned at best, and decadent at worst, as tastes changed and new, simpler cuisines came into vogue in Europe. At the same time, spices were eclipsed as exotic status symbols by new products such as tobacco, coffee, and tea. By solving the mystery of the spices' origins, the spice-seekers paradoxically devalued the treasure they had so arduously sought. Today most people walk past the spices in the supermarket, arrayed on shelves in

small glass bottles, without a second thought. In some ways it is a
sorry end to a once-mighty trade that reshaped the world.

LOCAL AND GLOBAL FOOD

Ideally suited as they were to long-distance freight, spices led to the
wiring up of the first global trade networks. The great distance they
traveled was one of the reasons people were prepared to pay so much
for them—some people, at least. But not everyone approved of
bringing these inessential, frivolous ingredients all that way: "For the
sake of this we go to India!" Pliny the Elder grumbled about pepper
in the first century A.D. Today a similar argument is advanced by
proponents of "local food," who advocate the consumption of foods
produced close to the consumer (within one hundred miles, say)
rather than shipped in from farther afield. They decry the trans-
portation of food that has, in some cases, traveled thousands of
miles from farm to plate; some local-food fundamentalists even try
to avoid nonlocal foods altogether. Pliny thought buying imported
food was simply a waste of money, but modern-day local-food advocates (or "locavores") generally make their case on environmental
grounds: Shipping all that food around causes carbon dioxide emissions that contribute to climate change. This has given rise to the
concept of "food miles"—the notion that the distance food is transported gives a reasonable measure of its environmental damage
caused, and that one should therefore eat local food to minimize
one's impact.

It sounds plausible enough, but the reality is rather more complex. For one thing, local products can sometimes have a greater environmental impact than those produced in other countries, simply
because some countries are better suited than others for production
of particular foods. Tomatoes are often grown in heated greenhouses
in Britain, for example, resulting in a larger volume of carbon emissions than tomatoes grown in Spain, even when the emissions produced by transporting Spanish tomatoes to Britain are included.

Similarly, a study carried out at Lincoln University in New Zealand found that lamb produced in that country produced far less carbon dioxide (563 kilograms per metric ton of meat) than lamb produced in Britain (2,849 kilograms per metric ton). This is largely because there is more room for pasture in New Zealand, so the lambs eat grass, whereas British lambs are given feed, the production of which is carbon-intensive. Shipping New Zealand lamb to Britain then incurred further emissions of 125 kilograms per metric ton, so that the "carbon footprint" of New Zealand lamb was much smaller even when transport was taken into account. It may be that the least polluting way to organize food production would be for countries or regions to concentrate on producing foods that can be made particularly efficiently given the local conditions, and to trade the resulting foods with each other.

Focusing on food's transport-related emissions may also be picking the wrong target. An American study found that transport accounted for 11 percent of the energy used in the food chain, compared with 26 percent for processing and 29 percent for cooking. In the case of potatoes, the emissions associated with cooking them far outweigh those involved in growing and transporting them. Whether or not you leave the lid on the pan when boiling your potatoes has more of an impact on the total carbon dioxide emissions than whether they were grown locally or far away. Another complicating factor is the wide variation in the efficiencies of different forms of transport. A large ship can carry a ton of food 800 miles on a gallon of fuel; the figures are about 200 miles for a train, 60 miles for a truck, and 20 miles for a car. So the drive to and from a shop or market can produce more emissions, for a given weight of food, than the whole of the rest of its journey.

Of course, not all the arguments made in favor of local food are environmental: There are social arguments, too. Local food can promote social cohesion, support local businesses, and encourage people to take more of an interest in where their food comes from and how it is grown. But there are also social arguments in favor of imported food. In particular, an exclusive focus on local foods would harm the prospects of farmers in developing countries who grow high-value crops for export to foreign markets. To argue that they should concentrate on growing staple foods for themselves, rather than more valuable crops for wealthy foreigners, is tantamount to denying them the opportunity of economic development.

There is undoubtedly some scope for "relocalization" of the food supply, and if nothing else, the food-miles debate is making consumers and companies pay more attention to food's environmental impact. But localism can be taken too far. Equating local food with virtuous food, today as in Roman times, is far too simplistic. The rich history of the spice trade reminds us that for centuries, people have appreciated exotic flavors from the other side of the world, and that meeting their needs brought into being a thriving network of commercial and cultural exchange. Hunter-gatherers were limited to local food by definition; but if subsequent generations had limited themselves in the same way, the world would be a very different place today. Admittedly, the legacy of the spice trade is mixed. The great spice-seeking voyages revealed the true geography of the planet and began a new epoch in human history. But it was also because of spices that European powers began grabbing footholds around the world and setting up trading posts and colonies. As well as sending Europeans on voyages of discovery and exploration, spices provided the seeds from which Europe's colonial empires grew.

PART IV

FOOD, ENERGY, AND
INDUSTRIALIZATION

7

NEW WORLD, NEW FOODS

The greatest service which can be rendered any country is to add a useful plant to its [agri]culture.

—THOMAS JEFFERSON

A Pineapple for the King

The portrait of King Charles II of England, painted around 1675, is not as simple as it looks. The king is shown wearing a knee-length coat and breeches, and standing in the elaborate gardens of a large house. Two spaniels attend him, and nearby kneels John Rose, the royal gardener, who is presenting Charles with a pineapple. The symbolism seems clear. At the time, pineapples were extremely rare in England, since they had to be imported from the West Indies and very few survived the voyage without spoiling. They were so valued that they were known as the "fruit of kings," a connotation strengthened by the leafy crown that adorns each pineapple. In England, the pineapple's association with kingly wealth and power dated back to 1661, when Charles had been sent one by a consortium of Barbados planters and merchants who wanted him to impose a minimum price on their main export, sugar. Charles received more than ten thousand petitions from various interest groups during the 1660s, so the gift of a pineapple, one of the first ever seen in England, was a clever move by the Barbados consortium that made their request stand out. It worked: Charles agreed to their proposal a few days after the pineapple's arrival.

The pineapple in the painting was more than simply a status symbol, however; it was also a reminder of England's rise as a maritime

Portrait of Charles II accepting a pineapple from John Rose.

trading power, and of its ascendancy in the West Indies in particular. Charles had passed the Navigation Acts during the 1660s, which
banned foreign ships from trading with English colonies and so encouraged a dramatic expansion of the English merchant fleet. In 1668
a pineapple had served as a reminder of England's growing naval
might at a banquet held by Charles in honor of the French ambassador, Charles Colbert. At the time, England and France were fighting
over colonial possessions in the West Indies, so the appearance of a
pineapple as the centerpiece of the dessert course emphasized the
king's commitment to his territories overseas. One observer at the
feast recorded that Charles cut the fruit up himself and offered pieces
of it from his own plate. This might sound like a gesture of humil-
ity, but was really a demonstration of his power: Only a king could
offer his guests pineapple.

    Lending further meaning to the painting was the fact that the
pineapple shown was an unusual fruit: It was, according to the
painting's title, "the first pineapple raised in England." It seems most
likely that the pineapple in question had been imported as a young

plant and had merely been ripened in England, rather than being grown from scratch—something that only became possible later, in the 1680s, with the invention of the heated greenhouse. Even so, to have ripened a tropical fruit in England was quite a feat, and it signaled the expertise of England's horticulturalists at a time when European nations were competing to discover, categorize, propagate, and exploit the wealth of plants from Asia and the Americas that had suddenly become available to them. In this new field of "economic botany," the pursuit of scientific knowledge went hand in hand with furthering the national interest, and botanical gardens were being established around the world as colonial laboratories.

The undisputed leaders in the field of economic botany in the late seventeenth century were the Dutch, who had pushed aside the Portuguese to become the dominant European power in the East at the time. The Dutch wanted to understand new plants for two main reasons: to find cures for the tropical diseases that were afflicting their sailors, merchants, and colonists; and to find new agricultural commodities, beyond the known spices, from which to make money. The Dutch set up botanical gardens at their colonial outposts at the Cape, at Malabar, Ceylon and Java, and in Brazil, all of which exchanged specimens with similar establishments back home, in Amsterdam and Leyden. These were much more ambitious than the botanical gardens established in Europe during the sixteenth century, starting in Italy in the 1540s, which had been chiefly medicinal in purpose. As England and France raced to emulate the Dutch and establish colonies and trading posts of their own, they also discovered an enthusiasm for economic botany. The history of the spice trade had shown that vast fortunes awaited anyone who could control the supply and trade of valuable foodstuffs; who knew what other plants were waiting to be exploited?

As if to emphasize the link between botanical and geopolitical mastery, some botanical gardens were even laid out to represent the world. Most were square, and were divided into four parts, one each for Europe, Africa, Asia, and the Americas. These areas were then further subdivided, right down to individual beds for particular plants. The botanists who established them dreamed of being able to gather the whole world's plants in one place. As the catalog of the Oxford Botanic Garden put it, "as all creatures were gathered into the Ark . . . so you have the plants of this world in microcosm in our garden." But this ambitious goal proved to be hopelessly unrealistic as the number of known plants mushroomed. The "Enquiry into Plants" by Theophrastus, an ancient Greek author, included only five hundred plants; the "Pinax Theatri Botanici," an epic work published by the Swiss botanist Caspar Bauhin in 1596, listed six thousand; and by the 1680s John Ray's "Historia Generalis Plantarum" listed more than eighteen thousand. In botany, as in so many other fields, the knowledge of the ancient authorities was found to be incomplete or plain wrong.

So the botanists served two masters: On the one hand they were members of an international research community, working together to add to mankind's understanding of nature, participants in a scientific revolution in which direct observation finally triumphed over received wisdom. On the other hand, they were expected to do their best to ensure that their own country would benefit the most from the new plants. Robert Kyd, a British army officer stationed in India who founded Calcutta Botanic Gardens in 1787, summed this up when he wrote that the gardens were established "not for the purpose of collecting rare plants as things of curiosity or furnishing articles for the gratification of luxury, but for establishing a stock for disseminating such articles as may prove beneficial to the inhabitants, as well as the natives of Great Britain, and which ultimately may tend to the extension of the national commerce and riches." Colonialism, commerce, and science went hand in hand; the number of plants a nation had at its disposal, and its botanists' ability to grow them outside their usual habitats, demonstrated that nation's technical prowess. Botany was regarded as the "big science" of its day, an indication of a country's might and sophistication, just as mastery of nuclear science or space technology is thought to be today. All this meant that the pineapple presented to Charles II was more than a mere fruit; it was a vivid symbol of his power.

As European explorers, colonists, botanists, and traders sought out new plants, learned how to nurture them, and worked out where else in the world they might also thrive, they reshaped the world's ecosystems. The "Columbian Exchange" of food crops between the Old and New worlds, in which wheat, sugar, rice, and bananas moved west and maize, potatoes, sweet potatoes, tomatoes, and chocolate moved east (to list just a handful of examples in each direction), was a big part of the story, but not the only part; Europeans also moved crops around within the Old and New worlds, transplanting Arabian coffee and Indian pepper to Indonesia, for example, and South American potatoes to North America. Of course, crops had always migrated from one place to another, but never with such speed, on such a scale, or over such large distances. The post-Columbian stirring of the global food pot amounted to the most significant reordering of the natural environment by mankind since the adoption of agriculture. New foods from foreign lands slotted into previously underexploited ecological niches, increasing the food supply in many cases. This was true of potatoes and maize in parts of Eurasia, peanuts in Africa and India, and bananas in the Caribbean, for example. Sometimes new crops were hardier than local ones: Sweet potatoes from the Americas caught on in Japan because they could survive the typhoons that occasionally destroyed the rice crop, and cassava, also from the Americas, was adopted in Africa after being found to be resistant to locusts, since its edible roots remain safely out of reach underground.

Despite the botanists' nationalist ambitions, attempts to monopolize new plants generally did not last long. Making money from sugar, for example, depended on having colonial possessions with the right climate, and that depended chiefly on military rather than botanical might. Even so, one European nation emerged as the winner of this colonial contest, though its victory took an entirely unexpected form. The exchange and redistribution of food crops remade the world, and in particular those parts of it around the Atlantic Ocean, in two stages. First, new foods and new trading patterns redefined the demographics of the Americas, Africa, and Europe. Having done so, they then contributed to Britain's emergence as the first industrialized nation. Had he known this in 1675, Charles II would no doubt have been proud, though he might have been disappointed to hear that the pineapple was not one of the many foods that would play a part in this tale. Instead, the two foods that are central to the story are sugar, which traveled west across the Atlantic, and the potato, which traveled in the opposite direction.

Columbus and his Exchange

The Columbian Exchange, as the historian Alfred Crosby has called it, was aptly named because it really did start with Christopher Columbus himself. Although many other people carried plants, animals, people, diseases, and ideas between the Old and New worlds in the years to follow, Columbus was directly responsible for two of the earliest and most important exchanges of food crops with the Americas. On November 2, 1492, having arrived at the island of Cuba, he sent two of his men, Rodrigo de Jerez and Luis de Torres, into the interior with two local guides. Columbus believed that Cuba was part of the Asian mainland, and he expected his men to find a large city where they could make contact with the emperor. Torres spoke a little Arabic, which would, it was assumed, be understood by the emperor's representatives. After four days the men returned, having failed to find either city or emperor. But they had, Columbus recorded, seen many fields of "a grain like millet that the Indians call maize. This grain has a very good taste when cooked, either roasted or ground and made into a gruel." This was the first time that Europeans had encountered maize, and Columbus probably took some back to Spain with him when he returned from his first voyage, in 1493; he certainly took back maize from his second expedition the following year.

Though maize was initially regarded as a botanical curiosity by European scholars, it soon became apparent that it was well suited to the southern Mediterranean climate and was, in fact, an extremely valuable crop. By the 1520s it had established itself in several parts of Spain and northern Portugal, and it soon afterward spread around the Mediterranean, into central Europe, and down the west coast of Africa. So rapid was the spread of maize around the world that its origins became obscured almost immediately. In Europe, it was variously known as Spanish corn, Indian corn, Guinea corn, and Turkey wheat, reflecting confusion about its provenance. And the speed with which maize reached China—it probably arrived there in the 1530s, though the first definite Chinese reference to it was not until 1555—led some people to the erroneous conclusion that maize must have been present in Europe and Asia before Columbus. Maize spread so quickly because it had such desirable properties. It grew well in soil that was too wet for wheat and too dry for rice, so it provided extra food from marginal land where existing Eurasian staples could not be grown. It also had a short growing season and produced a higher yield, per unit of land and labor, than any other grain. And whereas wheat typically produced four to six times as much grain per measure of seed sown, the figure for maize was between one hundred and two hundred.

If maize, the crop that Columbus took eastward, was a blessing, then sugarcane, the crop he took westward, was a curse. Having worked in his youth as a sugar buyer for Genoese merchants, Columbus was familiar with sugar cultivation. He realized that the new lands he had discovered were well suited to the production of this lucrative product, and he took sugarcane with him to Hispaniola on his second voyage to the Americas in 1493. If he could not find gold or spices, he could at least make sugar. Given the labor-intensive nature of its production, he would have to find sufficient manpower, of course. But Columbus had observed after his first voyage that "the

Indians have no weapons and are quite naked . . . they need only to be given orders to be made to work, to sow, or to do anything useful." In other words, he could put the locals to work as slaves.

Sugar and slavery had gone together for centuries. Sugarcane is originally from the Pacific islands, was encountered in India by the ancient Greeks, and was introduced to Europe by the Arabs, who began cultivating it on a large scale in the Mediterranean in the twelfth century using slaves from East Africa. Europeans acquired a taste for sugar during the Crusades and captured many of the Arab sugar plantations, which they manned with Syrian and Arab slaves. The slave-based production system was then transplanted to the Atlantic island of Madeira in the 1420s after its discovery by the Portuguese. During the 1440s the Portuguese increased sugar production by bringing in large numbers of black slaves from their new trading posts on the west coast of Africa. At first these slaves were kidnapped, but the Portuguese soon agreed to buy them, in return for European goods, from African slave-traders. By 1460 Madeira had become the world's largest sugar producer, and no wonder: It had an ideal climate for sugar, was close to the supply of slaves, and was on the edge of the known world, so that the brutal realities of sugar production were kept conveniently out of sight of the growing throng of European consumers. The Spanish, for their part, began making sugar on the nearby Canary Islands, again using slaves from Africa.

This proved to be merely the warm-up for what was to come in the New World. It was not until 1503 that the first sugar mill opened on Hispaniola. The Portuguese began production in Brazil around the same time, and the British, French, and Dutch established sugar plantations in the Caribbean during the seventeenth century. After attempts to enslave local people failed, chiefly because they succumbed to Old World diseases to which they had no immunity, the colonists began importing slaves directly from Africa. And so began the Atlantic slave trade. Over the course of four centuries, around eleven million slaves were transported from Africa to the New World, though this figure understates the full scale of the suffering, because as many as half of the slaves captured in the African interior died on the way to the coast. The vast majority of the slaves shipped across the Atlantic—around three quarters of them—were put to work making sugar, which became one of the main commodities in Atlantic trade.

This trade developed in the seventeenth and eighteenth centuries and ended up consisting of two overlapping triangles. In the first, commodities from the Americas, chief among them sugar, were shipped to Europe; finished goods, chiefly textiles, were shipped to Africa and used to purchase slaves; and those slaves were then shipped to the sugar plantations in the New World. The second triangle also depended on sugar. Molasses, the thick syrup left over from sugar production, was taken from the sugar islands to England's North American colonies, where it was distilled into rum. This rum was then shipped to Africa where, along with textiles, it was used as currency to buy slaves. The slaves were then sent to the Caribbean to make more sugar. And so on.

Having been an expensive luxury item at the time of the Crusades, sugar fell in price as production increased, and by the end of the eighteenth century it had become an everyday item for many Europeans. Demand grew as the exotic new drinks of tea, coffee, and cocoa (from China, Arabia, and the Americas, respectively) became popular in Europe, invariably sweetened with sugar. Having used fruit and honey as sweeteners for centuries, European consumers suddenly became accustomed to sugar, even addicted to it. The demand enriched Caribbean sugar barons, European merchants, and North American colonists. Rum became the most profitable manufactured item produced in New England, and by the early eighteenth century it accounted for 80 percent of exports. Attempts by the British government to restrict imports to New England of cheap molasses from the French sugar islands, in the form of the Sugar and Molasses Act of 1733 and the Sugar Act of 1764, were deeply unpopular with the colonists, causing the first of many disagreements and protests that ultimately led to the Declaration of Independence.

As well as being notable for its reliance on slavery and its economic importance, sugar production also crystallized a new model of industrial organization. Making sugar involved a series of processes: cutting the sugarcane, pressing it to extract the juice, boiling and skimming the juice, and then cooling it to allow the crystals of sugar to form, while the leftover molasses was distilled into rum. The desire to do all of this on a large scale, as quickly and efficiently as possible, led to the development of increasingly elaborate machinery and prompted the division of workers into teams that specialized in separate parts of the process.

In particular, sugar production depended on the use of rolling mills to press the cane. These could extract juice more efficiently than the old-fashioned methods of chopping up the stalks by hand and pounding it, or using screw presses. Rolling mills were also better suited to continuous production: Once pressed, the stalks could be used as fuel for the boilers in the next stage of the process. The machinery developed to process sugar—powered by wind, water, or animal power—was the most elaborate and costly industrial technology of its day, and it prefigured the equipment later used in the textile, steel, and paper industries.

Operating the rolling mills, tending the boiling cauldrons of juice, and working the distilling equipment could be dangerous, however. A moment's inattention when feeding sugar into the roller mill, or when handling the boiling sugar, could lead to horrific injuries or death. As one observer noted: "If a Boyler get any part into the scalding sugar, it sticks like Glew, or Birdlime, and 'tis hard to save either Limb or Life." Nobody would do such dangerous and repetitive work at the low salaries planters were offering, which is why the planters relied on slave labor. To minimize the risk of accidents, it made sense for workers to specialize in particular tasks. Even for less dangerous work, such as the cultivation of the cane, planters found that dividing their slaves into teams and giving them specific tasks made it easier to supervise their work and coordinate the different stages of the process.

*An engraving showing proto-industrial
sugar production in the West Indies.*

Establishing a sugar plantation required large capital investments to pay for land, buildings, machinery, and slaves. The resulting plantations were the largest privately owned businesses of their day, making their owners (who could expect annual profits of around 10 percent of capital invested) among the wealthiest men of the time. It has been suggested that profits from the sugar and slave trades provided the bulk of the working capital needed for Britain's subsequent industrialization. In fact, there is little evidence that this was the case. But the idea of organizing manufacturing as a continuous, production-line process, with powered, labor-saving machinery and workers specializing in particular tasks, does owe a clear debt to the sugar industry of the West Indies, where this arrangement first emerged on a large scale.

"LET THEM EAT POTATOES"

When Marie-Antoinette, the queen of France, heard that the peasants had no bread to eat, she is supposed to have declared, "Let them eat cake." In one version of the story, she said this when the starving poor clamored at her palace gates; in another, the queen made the remark while riding through Paris in her carriage and noting how ill-fed the people were. Or perhaps she said it when hungry mobs stormed the bakeries of Paris in 1775 and almost caused the post-ponement of the coronation of her husband, Louis XVI. In fact, she probably never said it at all. It is just one of many myths associated with the infamous queen, who was accused of all kinds of excess and debauchery by her political opponents in the run-up to the French Revolution in 1789. But the phrase encapsulates the perception of Marie-Antoinette as someone who professed to care about the starving poor but was utterly incapable of understanding their troubles. Even if she never advocated the substitution of cake for bread, how-ever, she did publicly endorse another foodstuff as a means of feeding the poor: the potato. She probably did not say "Let them eat potatoes" either, but that is what she and many other people thought. And it was not such a bad idea. In the late eighteenth cen-tury, potatoes were belatedly being hailed as a wonder food from the New World.

Europeans had first learned of potatoes in the 1530s, when the Spanish conquistadores embarked upon the conquest of the Inca Empire, which stretched right down the west coast of the South American continent. Potatoes were a mainstay of the Inca diet, alongside maize and beans. Originally domesticated in the region of Lake Titicaca, they then spread throughout the Andes and beyond. The Incas developed hundreds of varieties, each suited to a different combination of sun, soil, and moisture. But the value of potatoes was lost on the Europeans who first encountered them. The earliest written description, dating from 1537, describes them as "spherical roots which are sown and produce a stem with its branches and flow-ers, although few, of a soft purple color; and to the root of this same plant . . . they are attached under the earth, and are the size of an egg more or less, some round and some elongated; they are white and purple and yellow, floury roots of good flavor, a delicacy to the Indians and a dainty dish even for the Spaniards." Although a few potatoes were sent back to Spain, and spread from there to Europe's botanical gardens, they were not seized upon as a valuable new crop in the way that maize had been. By 1600 potatoes were being cultivated on a small scale in a few parts of Europe, since the Spanish had introduced them to their possessions in Italy and the Low Countries. In 1601 Clusius, a botanist in Leyden, described the potato and gave it the scientific name *Solanum tuberosum*. He noted that he had received specimens in 1588 and that potatoes were grown in Italy for consumption by both humans and animals.

Why did potatoes not prove more popular? After all, in the sandy soil of northern Europe they would eventually prove to be capable of producing two to four times as many calories per acre as had previously been possible with wheat, rye, or oats. Potatoes take only three to four months to mature, against ten for cereal grains, and can be grown on almost any kind of soil. One problem was that the first potatoes brought over from the Americas were adapted to growing in the Andes, where the length of the day does not vary much during the year. In Europe, where the length of the day varies far more, they initially produced a rather meager crop, and it took botanists a few years to breed new varieties that were well suited to the European climate.

But even then, Europeans were suspicious of this new vegetable. Unlike maize, which was recognizable as a previously unknown cousin of wheat and other cereal grains, potatoes were unfamiliar and alien. They were not mentioned in the Bible, which suggested that God had not meant men to eat them, said some clergymen. Their unaesthetic, misshaped appearance also put people off. To herbalists who believed that the appearance of a plant was an indication of the diseases it could cause or cure, potatoes resembled a leper's gnarled hands, and the idea that they caused leprosy became widespread. According to the second edition of John Gerard's *Herball*, published in 1633, "the Burgundians are forbidden to make use of these tubers, because they are assured that the eating of them causes leprosy." More scientifically inclined botanists took an interest in potatoes, the first known edible tubers, and identified them as members of the poisonous nightshade family. That did not help their reputation either: Potatoes came to be associated with witchcraft and devil worship.

At the beginning of the seventeenth century potatoes were widely regarded as suitable fodder for animals, but to be eaten by humans only as a last resort, when no other food was available. The potato made slow progress in the following years, being consumed only by the very rich (it was prized by some aristocratic gardeners and was served as a novelty) and the very poor (it became a staple food among the poor, first in Ireland, and then in parts of England, France, the Low Countries, the Rhineland, and Prussia). Famines brought the potato new converts, as people who had no choice but to eat potatoes soon discovered that they were not so terrible after all. One of the first acts of the Royal Society, Britain's pioneering scientific society, after its foundation in 1660, was to point out the value of potatoes in times of famine—on the basis that in years when the wheat crop failed, there was often a good potato harvest. But this advice was ignored, and it was only when famine struck, as it did in France in 1709, that the virtues of potatoes were made starkly clear and the threat of starvation forced people to put aside their prejudices.

A series of famines in the eighteenth century earned the potato some friends in high places. When the crops failed in 1740, Frederick the Great of Prussia urged wider cultivation of potatoes among his subjects. His government distributed a handbook explaining how to grow the new crop and distributed free seed potatoes. Other European governments did the same, making promotion of the potato official policy. In Russia, Catherine the Great's medical advisers convinced her that the potato could be an antidote to starvation;

governments in Bohemia and Hungary also advocated its cultivation. Sometimes potato advocacy was backed by force: Austrian peasants were threatened with forty lashes if they refused to embrace it. Warfare also helped to change attitudes. During their campaigns in northern Europe in the 1670s and 1680s, Louis XIV's armies encountered potatoes in Flanders and the Rhineland, where they were being grown in some quantity by this time. One observer noted that "the French Army found great support thereby by feeding the common Soldiers most plenteously; it is both delicious and wholesome."

Austrian, French, and Russian soldiers who fought in Prussia during the Seven Years' War (1756–63) saw how potatoes (planted at Frederick the Great's urging) sustained the local population, and they advocated their cultivation when they returned home. One advantage of the potato during wartime was that it remained hidden safely underground; even if an army camped on a field of potatoes, the farmer could still harvest them afterward.

One man's experience of potatoes during the Seven Years' War inspired him to become the potato's greatest champion. Antoine-Augustin Parmentier, a French scientist, served as a pharmacist in the French army. After being captured by the Prussians he spent three years in prison, and for much of that time he was given nothing more than potatoes to eat. He concluded that they were a nourishing and healthy food, and when the war ended and he returned to France he became a vocal potato advocate. After yet another poor harvest in 1770, when a prize was offered for the best essay on "foodstuffs capable of reducing the calamities of famine," Parmentier won with a eulogy to the potato. Even though potatoes were still widely believed to be poisonous and to cause disease, he won backing for his views in 1771 from the medical faculty at the Sorbonne university in Paris, which ruled that the potato was indeed fit for human consumption. Shortly afterward Parmentier published a detailed scientific analysis of the merits of the potato. But support among the scientific community was one thing; after years of effort,

Parmentier found that convincing people to cultivate and eat potatoes was quite another.

So he organized a series of publicity stunts. In 1785, at a banquet to celebrate the birthday of Louis XVI, Parmentier presented the king and queen with a bouquet of potato flowers, whereupon the king pinned one of the flowers to his lapel, and Marie-Antoinette put a garland in her hair. When the guests sat down to eat, several of the dishes included potatoes. With the endorsement of the king and queen, eating potatoes and wearing potato flowers soon became fashionable among the aristocracy. Parmentier also hosted several dinners of his own, serving potatoes prepared in a variety of ways to emphasize their versatility. (The American statesman and scientist Benjamin Franklin was among the celebrities invited to these dinners.) But Parmentier's greatest trick was to post armed guards around the fields just outside Paris, given to him by the king, where he was growing potatoes. This aroused the interest of the local people, who wondered what valuable crop could possibly require such security measures. Once the crop was ready, Parmentier ordered the guards to withdraw, and the locals duly rushed in and stole the potatoes. As hostility toward the potato finally crumbled, the king is said to have told Parmentier: "France will thank you some day for having found bread for the poor." But it was only some years later, after the French Revolution (during which Louis XVI and Marie-Antoinette were guillotined), that the king's prediction proved correct. In 1802 Napoleon Bonaparte instituted the order of the Legion d'Honneur, and Parmentier was among its first recipients. His service to the potato is remembered today in the form of several potato-based dishes that bear his name.

It was a similar if less poetic story elsewhere in Europe: The combination of famine, war, and government promotion meant that by 1800, the potato had established itself as an important new foodstuff. Sir Frederick Eden, an English writer and social researcher, wrote that in Lancashire "it is a constant standing dish, at every meal, breakfast excepted, at the tables of the Rich, as well as the

Poor . . . potatoes are perhaps as strong an instance of the extension of human enjoyment as can be mentioned." The potato was hailed as "the greatest blessing that the soil produces," "the miracle of agriculture," and "that most valuable of roots." After bad wheat harvests in 1793 and 1794, many people dropped their opposition to potatoes in 1795. That year the *Times* of London even printed recipes for potato soup and for bread with maize and potatoes. One factor that counted in the potato's favor was the high status of white bread, made from wheat, compared with brown bread, made from rye, oats, and barley. English workers who had become wealthy enough to switch from brown to white bread during the eighteenth century were very reluctant to switch back again. When times were hard, they would sooner eat potatoes.

In his book *The Wealth of Nations*, published in 1776, the Scottish philosopher and economist Adam Smith observed that "the food produced by a field of potatoes is not inferior in quantity to that produced by a field of rice, and much superior to what is produced by a field of wheat." Even allowing for the fact that potatoes contained a large amount of water, he noted, "an acre of potatoes will still produce six thousand weight of solid nourishment, three times the quantity produced by the acre of wheat." His praise of the potato continued with words that now seem prophetic: "Should this root ever become in any part of Europe, like rice in some rice countries, the common and favorite vegetable food of the people, so as to occupy the same proportion of the lands in tillage which wheat and other sorts of grain for human food do at present, the same quantity of cultivated land would maintain a much greater number of people, and . . . population would increase."

FROM COLUMBUS TO MALTHUS

Three centuries after Columbus's arrival in the Americas, the ensuing exchange of plants, diseases, and people had transformed the world's population and its distribution. Smallpox, chicken pox, influenza, typhus, measles, and other Old World diseases—many of them consequences of human proximity to domesticated animals such as pigs, cows, and chickens that had been unknown in the New World—had decimated the native peoples of the Americas, who lacked immunity to such diseases, paving the way for European conquest. Estimates of the size of the pre-Columbian population of the Americas vary from 9 million to 112 million, but a consensus figure of 50 million, which had been reduced by disease and warfare to some 8 million by 1650, gives an idea of the scale of the destruction. Even as their invisible biological allies wiped out the indigenous peoples of the Americas, Europeans began importing slaves from Africa on a vast scale to work on sugar plantations. The demographics of Africa and the Americas were transformed. But the Columbian Exchange also helped to alter the demographics of Eurasia.

In China, the arrival of maize and sweet potatoes contributed to the increase in population from 140 million in 1650 to 400 million in 1850. Since maize could be grown in areas that were too dry for rice, and on hillsides that could not be irrigated, it added to the food supply and allowed people to live in new places. The uplands of the Yangtze basin were deforested to make way for the production of indigo and jute, for example, and the peasants who grew them lived on maize and sweet potatoes, which grew well in the hills. Another practice that allowed food production to keep pace with a growing population was that of multiple cropping. When rice is grown in paddies, it absorbs most of its nutrients from water rather than soil, so it can be repeatedly cropped on the same land without the need to leave the land fallow to allow the soil to recover. Farmers in southern China could sometimes produce two or even three crops a year from a single plot of land.

In Europe, meanwhile, the new crops played a part in enabling the population to grow from 103 million in 1650 to 274 million in 1850. During the sixteenth century, Europe's staple crops, wheat and rye, produced about half as much food per hectare (measured by weight) as maize did in the Americas, and about a quarter as much as rice did in southern Asia. So the arrival of maize and potatoes in Europe provided a way to produce much more food from the same amount of land. The most striking example was that of Ireland, where the population increased from around 500,000 in 1660 to 9 million in 1840—something that would not have been possible without the potato. Without it, the whole country could only have produced enough wheat to support 5 million people. Potatoes meant that there was enough food to support nearly twice this number, even as wheat continued to be grown for export. Potatoes could be grown on European land that was unsuitable for wheat, and were far more reliable. Being better fed made people healthier and more resistant to disease, causing the death rate to fall and the birth rate to rise. And what potatoes did in the north of Europe, maize did in the south: the populations of Spain and Italy almost doubled during the eighteenth century.

As well as adopting the new crops, European farmers increased production by bringing more land under cultivation and developing new agricultural techniques. In particular, they introduced crop rotations involving clover and turnips (most famously, in Britain, the "Norfolk four-course rotation" of turnips, barley, clover, and wheat). Turnips were grown on land that would otherwise have been left fallow, and then fed to animals, whose manure enhanced the barley yields the following year. Feeding animals with turnips also meant that land used for pasture could instead be used to grow crops for human consumption. Similarly, growing clover helped to restore the fertility of the soil to ensure a good wheat harvest in the following year. Another innovation was the adoption of the seed drill, a horse-drawn device which placed seeds into holes in the soil at a precise depth. Sowing seeds in this way, rather than scattering them in the traditional manner, meant that crops were properly spaced in neat rows, making weeding easier and ensuring that adjacent plants did not compete for nutrients. Again, this helped to increase the yields of cereal crops.

By the end of the eighteenth century, however, there were signs that the European surge in agricultural productivity could no longer keep up with population growth. The problem was most noticeable in England, which had been more successful than other European countries in increasing its food production, and so had more difficulty maintaining the pace it had set itself once the population expanded. During the first half of the century, England had exported grain to continental Europe; but after 1750 the growing population, and a succession of bad harvests, led to shortages and higher prices. Agricultural output was still growing (by around 0.5 percent a year), but only at about half the rate of population growth (around 1 percent a year), so the amount of food per head was falling. The same thing was happening across Europe: anthropometric research shows that European adults born between 1770 and 1820 were, on average, noticeably shorter than previous generations had been.

In China, rice production could be increased using more labor and more multiple cropping. But that was not an option for European crops, so the obvious thing to do was to bring even more land under cultivation. The problem was that the supply of land was finite, and it was needed for other things besides agriculture: to grow wood for construction and fuel, and to accommodate Europe's growing cities. Again, the problem was particularly acute in England, where urbanization had been most rapid. People began to worry that the population would soon outstrip the food supply. The problem was elegantly summarized by the English economist Thomas Malthus, who published *An Essay on the Principle of Population* in 1798. It was an extraordinarily influential work, and its main argument runs as follows:

The power of population is indefinitely greater than the power in the earth to produce subsistence for man. Population, when unchecked, increases in a geometrical ratio. Subsistence increases only in an arithmetical ratio. A slight acquaintance with numbers will shew the immensity of the first power in comparison of the second. By that law of our nature which makes food necessary to the life of man, the effects of these two unequal powers must be kept equal. This implies a strong and constantly operating check on population from the difficulty of subsistence. This difficulty must fall somewhere and must necessarily be severely felt by a large portion of mankind.

Malthus thought that this predicament, which is now known as a "Malthusian trap," was inescapable. Given the chance, the population would double every twenty-five years or so, and then double again after the same interval, increasing in a geometric ratio; and despite the rapid increase in agricultural productivity of the preceding decades it was difficult to see how food production could possibly keep up. Even if food production could somehow be doubled from its level in the 1790s, that would only buy another twenty-five years' breathing space; it was hard to imagine how it could be doubled again. "During the next period of doubling, where will the food be found to satisfy the importunate demands of the increasing numbers?" Malthus asked. "Where is the fresh land to turn up?" Rapid population growth had, Malthus noted, been possible in the North American colonies, but that was because the population was relatively small in relation to the abundant land available.

"I see no way by which man can escape from the weight of this law which pervades all animated nature," he gloomily concluded. "No fancied equality, no agrarian regulations in their utmost extent, could remove the pressure of it even for a single century. And it appears, therefore, to be decisive against the possible existence of a society, all the members of which should live in ease, happiness, and comparative leisure; and feel no anxiety about providing the means of subsistence for themselves and families." He anticipated a future of food shortages, starvation, and misery. The potato, Malthus believed, was partly to blame. Having been championed as a remedy for starvation, it now seemed to be hastening the onset of an apparently inevitable crisis. And even if it provided enough food to go around, Malthus argued, the potato caused the population to increase far beyond the opportunities for employment. With hindsight, of course, we can appreciate the irony that Malthus pointed out the biological constraints on population and economic growth just at the moment when Britain was about to demonstrate, for the first time in human history, that they no longer applied.

8

THE STEAM ENGINE AND THE POTATO

It is the fashion to extol potatoes, and to eat potatoes. Every one joins in extolling potatoes, and all the world like potatoes, or pretend to like them, which is the same thing in effect.

—WILLIAM COBBETT, ENGLISH FARMER AND PAMPHLETEER, 1818

"THE OFFSPRING OF AGRICULTURE"

From the dawn of prehistory to the beginning of the nineteenth century, almost all of the necessities of life had been provided by things that grew on the land. The land supplied food crops of various kinds; wood for fuel and construction; fibers with which to make clothing; and fodder for animals, which in turn provided more food, along with other useful materials such as wool and leather. Butchers, bakers, shoemakers, weavers, carpenters, and shipbuilders depended on animal or vegetable raw materials, all of which were the products, directly or indirectly, of photosynthesis—the capture of the sun's energy by growing plants. Since all these things came from the land, and since the supply of land was limited, Thomas Malthus concluded that there was an ecological limit that growing populations and economies would eventually run into. He first made this prediction on the eve of the nineteenth century, and he refined his argument in the following years.

Yet Britain did not hit the ecological wall that Malthus anticipated. Instead, it vaulted over it and broke free of the constraints of the "biological old regime" in which everything was derived from the produce of the land. Rather than growing most of its own food,

Britain concentrated on manfacturing industrial goods, notably cotton textiles, which could then be traded for food from overseas. During the nineteenth century the population more than tripled, but the economy grew faster still, so that the average standard of living increased—an outcome that would have astonished Malthus. Britain had dealt with the looming shortage of food by reorganizing its economy. By switching from agriculture to manufacturing, Britain became the first industrialized nation in the world.

To be fair, Malthus could hardly have been expected to see this coming, since nothing like it had ever happened before. And none of it was planned: It was the accidental result of the convergence of several independent trends. Three of the most important related to changes in food production: greater specialization in handicrafts, prompted by rising agricultural productivity; the growing use of fossil fuels, initially as a land-saving measure; and an increasing emphasis on importing rather than growing food.

The first step along the road from a farm-based to a factory-based economy was the growth of rural industry, in the form of home-based manufacturing and handicrafts. This happened throughout Europe, but it was particularly notable in England because of the unusually rapid growth in English agricultural productivity. By 1800 only 40 percent of the male labor force worked on the land, compared with 65 to 80 percent in continental Europe. The number of men working in agriculture in 1800 was about the same as it had been two hundred years earlier, but the introduction of new crops and improved farming techniques meant that each one was producing twice as much food. This high productivity liberated ever more workers from the land and prompted people to move into rural manufacturing, as Adam Smith explained:

> An inland country naturally fertile and easily cultivated produces a great surplus of provisions beyond what is necessary for maintaining the cultivators . . . Abundance, therefore, renders provisions cheap, and encourages a great number of workmen to settle

in the neighbourhood, who find that their industry there can procure them more of the necessities and conveniences of life than in other places. They work up the material of manufacture which the land produces, and exchange their finished work, or what is the same thing the price of it, for more materials and provisions. They give a new value to the surplus part of the rude produce . . . and they furnish the cultivators with something in exchange for it that is either useful or agreeable to them. The cultivators get a better price for their surplus produce, and can purchase cheaper other conveniences which they have occasion for . . . The manufacturers first supply the neighbourhood, and afterwards, as their work improves and refines, more distant markets . . . In this manner have grown up naturally the manufactures of Leeds, Halifax, Sheffield, Birmingham and Wolverhampton. Such manufactures are the offspring of agriculture.

Once rural manufacturing had established itself in England, it intensified in the northern half of the country during the eighteenth century in response to the adoption of new agricultural techniques in the south. The use of clover and turnips in rotation with wheat and barley to increase cereal yields was less efficient on the heavy clay soils of the north and west of England, so people in those regions concentrated instead on livestock farming and manufacturing, and used the proceeds to buy grain from the south of the country. The result, by chance, was a concentration of manufacturing in just the regions of England where there were rich deposits of coal.

The Fuels of Industry

The shift to using coal rather than wood as a fuel was a second trend that contributed to Britain's industrialization. People much preferred burning wood rather than coal in their homes, but as land became more sought after for agricultural use, areas that had previously provided firewood were cleared to make way for farming. The price of firewood shot up—it increased threefold in western European cities between 1700 and 1800—and people turned to coal as a cheaper fuel. (It was cheap in England, at least, since there were plentiful deposits near the surface.) One ton of coal provides the same amount of heat as the wood that can be sustainably harvested each year from one acre of land. In England and Wales, some seven million acres of land that had previously provided wood, or around one fifth of the total surface area, were taken under cultivation between 1700 and 1800. This ensured that the growth of the food supply could continue to keep pace with the population—but required everybody to switch to burning coal.

And switch they did: The actual consumption of coal by 1800 was about ten million tons a year, providing as much energy as would otherwise have required ten million acres to be set aside for fuel production. At this point Britain accounted for 90 percent of world coal output, by some estimates. When it came to fuel, at least, Britain had already escaped from the constraints of the biological old regime. Rather than relying on living plants to trap sunlight to produce fuel, coal provided a way to tap vast reserves of past sunlight, accumulated millions of years ago and stored underground in the form of dead plants.

Although it was originally exploited as an alternative to wood for domestic heating, the abundance of coal meant that it was soon being put to other uses. Arthur Young, an English agricultural writer and social observer, was struck by the relative scarcity of glass in windows while traveling in France in the 1780s; it was far more widespread in England by this time because coal provided cheap energy for glassmaking. (French glassmakers, meanwhile, were so desperate for fuel that they had resorted to burning olive pits.) Coal was also heavily used by the textile industry, to warm the liquids used in bleaching, dyeing, and printing and to heat drying rooms and presses. Coal enabled a rapid expansion in the production of iron and steel, which had previously been smelted using wood. And, of course, coal was used to power steam engines, a technology that emerged from the coal industry itself.

Once England's outcropping surface deposits of coal had been depleted, it was necessary to sink mine shafts, and to ever greater depths—but the deeper they went, the more likely they were to flood with water. The steam engine invented by Thomas Newcomen in 1712, building on the work of previous experimenters, was built specifically to pump water out of flooded mines. Early steam engines were very inefficient, but this did not matter very much since they were powered by coal—and in a coal mine the fuel was, in effect, free. Hundreds of Newcomen engines had been installed in mines around England by 1800. The next step was taken by James Watt, a Scottish inventor who was asked to repair a Newcomen engine in 1763 and quickly realized how its wasteful design could be improved upon. His design, completed in 1775, was much more efficient and was also better suited to driving machinery.

This meant steam power could be applied to the various labor-saving devices that had been devised in the textile industry, providing an enormous increase in productivity. In 1790 the first steam-powered version of Samuel Crompton's "mule," a machine that spun cotton into yarn, increased the output of thread per worker 100-fold over a manual spinning wheel, for example. So much thread could be produced that looms also had to be automated to make use of it. By putting these various machines together in a single factory, so that the product of one stage of processing could be passed on to the next stage, as on a sugar plantation, it was possible to achieve further improvements in productivity. By the end of the eighteenth century Britain could produce textiles so cheaply and in such abundance that it began exporting them to India, devastating that country's traditional weaving trade in the process.

The third shift that underpinned Britain's Industrial Revolution was a far greater reliance on food imports. Just as it used coal from underground to power its new steam engines, Britain used food from overseas to provide energy for its workers. From its possessions in the West Indies, it brought in vast quantities of sugar, which provided an astonishing proportion of Britain's caloric intake during the nineteenth century, increasing from 4 percent of all calories consumed in 1800 to 22 percent by 1900. Sugar flowed eastward across the Atlantic, paying for manufactured goods that traveled in the opposite direction. Since an acre of sugar produces as many calories as nine to twelve acres of wheat, imported sugar provided the caloric equivalent of the produce of 1.3 million "ghost acres" of wheat-farming land in 1800, rising to 2.5 million acres in 1830 and around 20 million acres by 1900. Britain had clearly escaped the constraints of its limited land area by producing industrial goods, which did not require much land to manufacture, and trading them for food, which did.

Sugar was of course used to sweeten tea, the favored drink of industrial workers, which helpfully delivered energy (from the sugar) and kept them alert during long shifts (since tea contains caffeine). Sugar was also consumed as a foodstuff, to enliven an otherwise monotonous diet: It could be added to porridge in the form of treacle or molasses, and eaten as jam (containing 50 to 65 percent sugar) in sandwiches. Treacle or jam spread on bread was favored by working families in the industrial cities because it was a cheap source of calories and could be prepared quickly without the need to cook anything. Many women were now working in factories, and they no longer had time to prepare soup. The price of sugar fell and the availability of jam shot up after 1874, when Britain abolished its tariffs on sugar imports, which dated all the way back to Charles II and his pineapple in 1661.

It was not just the sugar in the jam that was imported; so too, increasingly, was the wheat used to make the bread. As the prospect of food shortages loomed in the late eighteenth century, Britain began to import more food from Ireland. Following the Act of Union of 1801, Ireland was technically part of the United Kingdom, but in practice it was treated as an agricultural colony by the English. Laws which had forbidden the importing of Irish animal products into England had been repealed in 1766, and by the end of the 18th century imports of Irish beef had gone up threefold, butter sixfold, and pork sevenfold. By the early 1840s, imports from Ireland were supplying one sixth of England's food. This food was produced by men who worked on the best, most easily cultivated land and were typically given small patches of inferior land on which they grew potatoes to support themselves and their families. The English could only keep eating bread, in short, because the Irish were eating potatoes. By sustaining Irish farm workers, the potato helped to fuel the first few decades of British industrialization.

The Potato Famine and Its Consequences

Britain's example appeared to have proved Malthus wrong, but in at least one respect he was ominously prescient. At the beginning of the nineteenth century Malthus had disagreed with the idea that potatoes provided the answer to the food problem, as they seemed to have done in Ireland. In *The Question of Scarcity Plainly Stated and Remedies Considered*, published in 1800, Arthur Young had suggested that the British government ought to give every country laborer with three or more children half an acre of land on which to grow potatoes and keep one or two cows. "If each had his ample potato-ground and a cow, the price of wheat would be of little more consequence to them than it is to their brethren in Ireland," he wrote. But Ireland's reliance on the potato was not something that other countries should seek to emulate, Malthus declared. For if people became dependent on potatoes, a failure of the potato crop would be a catastrophe. "Is it not possible," he wrote in response to Young's proposal, "that one day the potato crop itself may fail?"

Just such a catastrophe struck Ireland in the autumn of 1845. In retrospect it was a disaster waiting to happen. The potato crop had failed in previous years, at least in some parts of Ireland, and there had been a run of bad years in the 1830s. But the crop failure of 1845, caused by a previously unknown disease, was on an entirely different scale, and affected the whole country. The potato plants started to wither, while underground the tubers began to rot; fields full of apparently healthy plants were reduced to black, devastated foliage within days. This was the potato blight, caused by *Phytophthora infestans*, a fungus from the New World that crossed the Atlantic for the first time in 1845. Even potatoes that had been dug up before the blight manifested itself went bad and rotten within a month. What was expected to be a bumper crop—2.5 million acres of potatoes had been planted, 6 percent more than the previous year—was instead a total loss.

The scale of the devastation was unlike anything seen in some parts of Europe since the Black Death. The potato crop failed again in 1846, and the famine continued because farmers gave up planting potatoes in subsequent years. The people faced not just starvation, but disease. William Forster, a Quaker who visited Ireland in January 1847, recalled the scene in one village:

> The distress was far beyond my powers of description. I was quickly surrounded by a mob of men and women, more like famished dogs than fellow creatures, whose figures, looks and cries, all showed that they were suffering the ravening agony of hunger . . . in one [cabin] there were two emaciated men, lying at full length, on the damp floor . . . too weak to move, actually worn down to skin and bone. In another a young man was dying of dysentry; his mother had pawned everything . . . to keep him alive; and I never shall forget the resigned, uncomplaining tone in which he told me that all the medicine he wanted was food.

In Ireland around one million people starved to death as a result of the famine or were carried off by the diseases that spread in its wake. Another million emigrated to escape the famine, many of them to the United States. The potato blight also spread across Europe, and for two years there were no potatoes to be had anywhere. But Ireland's unrivaled dependence on the potato meant that it suffered the most.

As the magnitude of the disaster became apparent in late 1845, the British prime minister, Sir Robert Peel, found himself in a difficult situation. The obvious response to the famine was to import grain from abroad to relieve the situation in Ireland. The problem was that such imports were at the time subject by law to a heavy import duty to ensure that homegrown grain would always cost less, thus protecting domestic producers from cheap imports. The Corn Laws, as they were known, were at the heart of a long-running debate that had pitted the aristocratic landowners, who wanted the laws to stay in place, against an alliance of opponents led by industrialists, who demanded their abolition.

The landowners argued that it was better to rely on homegrown wheat than unreliable foreign imports, and warned that farmers would lose their jobs; they left unspoken their real concern, which was that competition from cheap imports would force them to reduce the rents they charged the farmers who worked their land. The industrialists said it was unfair to keep the price of wheat (and hence bread) artificially high, given that most people now bought food rather than growing their own; but they also knew that abolition would reduce demands for higher wages, since food prices would fall. Industrialists also hoped that cheaper food would leave people with more money to spend on manufactured goods. And they favored abolition of the Corn Laws because it would advance the cause of "free trade" in general, ensuring easy access to imported raw materials on one hand, and export markets for manufactured goods on the other. The debate over the Corn Laws was, in short, a microcosm of the much larger fights between agriculture and industry, protectionism and free trade. Was Britain a nation of farmers or industrialists? Since the landowners controlled Parliament, the argument had raged throughout the 1820s and 1830s to little effect.

The outcome was determined by the potato, as the famine in Ireland brought matters to a head. Peel, who had vigorously opposed the abolition of the Corn Laws in a Parliamentary debate in June 1845, realized that suspending the tariff on imports to Ireland in order to relieve the famine, but keeping it in place elsewhere, would cause massive unrest in England, where people would still have to pay artificially high prices. He became convinced that there was no alternative but to abolish the Corn Laws altogether, a reversal of his government's policy. At first he was unable to persuade his political colleagues, but some of them changed their minds as the news from Ireland worsened and it became apparent that the survival of the government itself was at stake. Finally, with a vote in May 1846, the Corn Laws were repealed. The support of the Duke of Wellington, an aristocratic war hero who had long been a strong supporter of the Corn Laws, was crucial. He persuaded the landowners who sat in the House of Lords to back the repeal on the grounds that the survival of the government was more important. But he privately conceded that "those damned rotten potatoes" were to blame for the demise of the Corn Laws.

The lifting of the tariff on imported grain opened the way for imports of maize from America, though in the event the government mishandled the aid effort and it made little difference to the situation in Ireland. The removal of the tariff also meant that wheat could be imported from continental Europe to replace the much diminished Irish supply. In the second half of the nineteenth century British wheat imports soared, particularly once the construction of railways in the United States made it easy to transport wheat from the Great Plains to the ports of the East Coast. Within Britain, meanwhile, the shift from agriculture to industry accelerated. The area of land under cultivation and the size of the agricultural workforce both went into decline in the 1870s. By 1900, 80 percent of Britain's main staple, wheat, was being imported, and the proportion of the labor force involved in agriculture had fallen to less than 10 percent.

Coal was not the only fuel that had driven this industrial revolution. The growth in agricultural productivity that had started two centuries earlier (supplemented by sugar from the Caribbean) and the supply of wheat from Ireland (made possible by the potato) had also played their part in carrying England over the threshold into the new industrial age. And by clearing away the obstacle to a greater reliance on food imports, the tragedy of the potato famine helped to complete the transformation.

FOOD AND ENERGY REVISITED

It is no exaggeration to suggest that the Industrial Revolution marked the beginning of a new phase in human existence, just as the Neolithic revolution associated with the adoption of farming had done some ten thousand years earlier. Both were energy revolutions: Deliberate farming of domesticated crops made a greater proportion of the solar radiation that reaches Earth available to mankind, and the Industrial Revolution went a step farther, exploiting solar radiation from the past, too. Both caused massive social changes: a switch from hunting and gathering to farming in the former case, and from agriculture to industry in the latter. Both took a long time to play out: It was thousands of years before farmers outnumbered hunter-gatherers globally, and industrialization has only been under way for 250 years, so only a minority of the world's population lives in industrialized countries so far—though the rapid development of China and India will soon tip the balance. And both are controversial: Just as it is possible to argue that hunter-gatherers were better off than farmers and that the adoption of agriculture was a big mistake, a case can also be made that industrialization has caused more problems than it has solved (though this argument is most often advanced by disillusioned people in rich, industrialized countries). There have been dramatic environmental consequences in both cases, too: Agriculture led to widespread deforestation, and industrialization has produced vast quantities of carbon dioxide and other greenhouse gases that have started to affect the world's climate.

In this sense the industrialized countries have not escaped Malthus's trap after all, but have merely exchanged one crisis, in which the limiting factor was agricultural land, for another, in which the limiting factor is the atmosphere's ability to absorb carbon dioxide. The possibility that the switch to fossil fuels might provide only a temporary respite from Malthusian pressures occurred even to nineteenth-century writers, notably William Stanley Jevons, an English economist and author of *The Coal Question*, published in 1865. "For the present," he wrote, "our cheap supplies of coal and our skill in its employment, and the freedom of our commerce with other wider lands, render us independent of the limited agricultural area of these islands, and apparently take us out of the scope of Malthus's doctrine." The word *apparently* did not appear in the first edition of the book, but Jevons added it to a later edition shortly before his death in 1882.

He was right to worry. In the early twenty-first century, renewed concerns about the connection between energy supplies and the availability of sufficient land for food production have been raised once again by the growing enthusiasm for biofuels, such as ethanol made from maize and biodiesel made from palm oil. Making fuel from such crops is appealing because it is a renewable source of energy (you can grow more next year) and over its life cycle it can produce fewer carbon emissions than fossil fuels. As plants grow, they absorb carbon dioxide from the air; they are then processed into biofuel, and the carbon dioxide goes back into the atmosphere when the fuel is burned. The whole process would be carbon neutral, were it not for the emissions associated with growing the crops in the first place (fertilizer, fuel for tractors, and so on) and then processing them into biofuels (something that usually requires a lot of heat). But exactly how much energy is required to produce various biofuels, and the level of associated carbon emissions, varies from crop to crop. So some biofuels make more sense than others.

The type that makes least sense is ethanol made from maize (corn), which is, unfortunately, the predominant form of biofuel, accounting for 40 percent of world production in 2007, most of it in the United States. The best-guess figures suggest that burning a gallon of corn ethanol produces only about 30 percent more energy than was needed to produce it, and reduces greenhouse-gas emissions by about 13 percent compared with conventional fossil fuel.

That may sound impressive, but the corresponding figures for Brazilian sugarcane ethanol are about 700 percent and 85 percent respectively; for biodiesel made in Germany they are 150 percent and 50 percent. Put another way, making a gallon of corn ethanol requires four fifths of a gallon of fossil fuel (not to mention hundreds of gallons of water), and does not reduce greenhouse-gas emissions by very much. America's corn-ethanol drive makes even less sense on economic grounds: To achieve these meager reductions in emissions, the United States government subsidizes corn-ethanol production to the tune of some seven billion dollars a year, and also imposes a tariff on sugarcane ethanol from Brazil to discourage imports. Corn ethanol seems to be an elaborate scheme to justify farming subsidies, rather than a serious effort to reduce greenhouse-gas emissions. England abolished its farmer-friendly Corn Laws in 1846, but America has just introduced new ones.

Enthusiasm for corn ethanol and other biofuels is one of the factors that has helped to drive up food prices as crops are diverted to make into fuel, so that they are in effect fed to cars, not people. Opponents of biofuels like to point out that the maize needed to fill a vehicle's twenty-five-gallon tank with ethanol would be enough to feed one person for a year. Since maize is also used as an animal feed, its higher price makes meat and milk more expensive, too. And as farmers switch their land from growing other crops to growing corn instead, those other crops (such as soy) become scarcer, and their prices also rise. Food and fuel are, it seems, once again competing for agricultural land. Cheap coal meant that English landowners in the eighteenth century realized their land was more valuable for growing food than fuel; concern about expensive oil today means American farmers are making the opposite choice, and growing crops for fuel rather than for food.

Biofuels need not always compete with food production, however. In some cases, it may be possible to grow biofuel feedstocks on marginal land that is unsuitable for other forms of agriculture. And those feedstocks need not be food crops. One potentially promising approach is that of cellulosic ethanol, in which ethanol is made from fast-growing, woody shrubs, or even from trees. In theory, this would be several times more energy efficient even than sugarcane ethanol, could reduce greenhouse-gas emissions by almost as much (a reduction of around 70 percent compared with fossil fuels), and would not encroach upon agricultural land. The problem is that the field is still immature, and expensive enzymes are needed to break down the cellulose into a form that can be made into ethanol. Another approach involves making biofuel from algae, but again the technology is still in its early days.

What is clear is that the use of food crops for fuel is a step backward. The next logical step forward, after the Neolithic and Industrial revolutions, is surely to find new ways to harness solar energy beyond growing crops or digging up fossil fuels. Solar panels and wind turbines are the most obvious examples, but it may also be possible to tinker with the biological mechanism of photosynthesis to produce more efficient solar cells, or to create genetically engineered microbes capable of churning out biofuels. The trade-off between food and fuel has resurfaced in the present, but it belongs in the past.

PART V

Food as a
Weapon

9

THE FUEL OF WAR

Amateurs talk tactics, but professionals talk logistics.

—ANONYMOUS

The fate of Europe and all further calculations depend upon the question of food. If only I have bread, it will be child's play to beat the Russians.

—NAPOLEON BONAPARTE

"MORE SAVAGE THAN THE SWORD"

What is the most devastating and effective weapon in the history of warfare? It is not the sword, the machine gun, the tank, or the atom bomb. Another weapon has killed far more people and determined the outcomes of numerous conflicts. It is something so obvious that it is easy to overlook: food, or more precisely, control of the food supply. Food's power as a weapon has been acknowledged since ancient times. "Starvation destroys an army more often than does battle, and hunger is more savage than the sword," noted Vegetius, a Roman military writer who lived in the fourth century A.D. He quoted a military maxim that "whoever does not provide for food and other necessities, is conquered without fighting."

For most of human history, food was literally the fuel of war. In the era before firearms, when armies consisted of soldiers carrying swords, spears, and shields, food sustained them on the march and gave them the energy to wield their weapons in battle. Food, including fodder for animals, was in effect both ammunition and fuel.

Maintaining the supply of food was therefore critical to military success; a lack of food, or its denial by the enemy, would lead swiftly to defeat. Before the advent of mechanized transport, keeping an army supplied with food and fodder often imposed significant constraints on where and when it could fight, and on how fast it could move. Although other aspects of warfare changed dramatically from ancient times to the Napoleonic era, the constraints imposed by food persisted. Soldiers could only carry a few days' worth of supplies on their backs; using pack animals or carts allowed an army to carry more supplies and equipment, but fodder for the animals was then needed, and the army's speed and mobility suffered.

This was recognized in the fourth century B.C. by Philip II of Macedonia, who introduced a number of reforms that were extended by his son, Alexander, to create the fastest, lightest, and most agile force of its day. Families, servants, and other followers, who sometimes equalled the soldiers in number, were restricted to an absolute minimum, allowing the army to throw off its immense tail of slow-moving people and carts. Soldiers were also required to carry much of their own equipment and supplies, with pack animals rather than carts carrying the rest. With fewer animals there was less need to find fodder, and the army became more mobile, particularly over difficult terrain. All this gave Alexander's army a clear advantage, allowing him to launch lightning strikes that struck fear into his enemies, according to Greek historians. Satibarzanes, a Persian governor, "learning of Alexander's proximity and astounded at the swiftness of his approach, fled with a few Arian horsemen." The Uxians, a Persian hill tribe, were "astounded by Alexander's swiftness, and fled without so much as coming to close quarters." And Bessus, a treacherous Persian nobleman, was "greatly terrified by Alexander's speed." Alexander's mastery of the mechanics of supplying his army—a field known today as logistics—enabled him to mount one of the longest and most successful military campaigns in history, conquering a swath of territory from Greece to the Himalayas.

Armies in history rarely brought along all of their own food supplies, however, and Alexander's was no exception. Food and fodder were also drawn from the surrounding country as the soldiers marched through. Such foraging could be an efficient way to feed an army, but it had the disadvantage that if the soldiers stopped moving, the local area would be rapidly depleted. Initially the army would have plenty of food at its disposal, but on each successive day foraging parties would have to travel farther to reach areas that had not yet been stripped of food. Alexander's rule of thumb, which was still valid centuries later, was that an army could only forage within a four-day radius of its camp, because a pack animal devours its own load within eight days. An animal that travels four days through barren country to gather food must carry four days' worth of food for its outward journey; it can then load up with eight days' worth of food, but will consume half of this on the return journey, leaving four days' worth—in other words, the amount it started off with. The length of time an army could stay in one place therefore depended on the richness of the surrounding country, which in turn depended on the population density (more people would generally have more food that could be appropriated) and the time of year (there would be plenty of food available just after the harvest, and very little available just before it). Alexander and other generals had to take these factors into account when choosing the routes of marches and the timing of campaigns.

Delivering supplies in bulk to an army on campaign was best done by ship, which was the only way to move large quantities of food quickly in the ancient world. Pack animals or carts could then carry supplies the last few miles from the port to the army's inland bases when necessary. This compelled armies to operate relatively close to rivers and coasts. As Alexander conquered the lands around the Mediterranean he was able to rely on his fleet to deliver supplies, provided his soldiers secured the ports along the coast beforehand. Moving from port to port, the soldiers carried a few days' worth of supplies and supplemented them by living off the land when possible. In the centuries after Alexander's death, the Romans took his logistic prowess a stage further. They established a network of roads and supply depots throughout their territory to ensure that supplies could be moved quickly and in quantity when needed. Their depots were resupplied by ship, which made it difficult for Roman armies to operate more than seventy-five miles from a coast or a large river. This helps to explain why Rome conquered the lands around the Mediterranean, and why the northern boundaries of its territory were defined by rivers. Maintaining permanent supply depots meant that a large force could move quickly through Roman territory without having to worry about finding food or fodder. The Roman army also introduced rules to govern the process of foraging while on campaign.

In enemy territory, demanding food requisitions from the surrounding area served two purposes: It fed the invading army and impoverished the local community. Food in such situations was literally a weapon: A marauding army could strip a region bare and cause immense hardship. As a medieval Chinese military handbook puts it, "If you occupy your enemies' storehouses and granaries and seize his accumulated resources in order to provision your army continuously, you will be victorious." Sometimes merely the threat of seizure was enough. In Alexander's case, local officials often surrendered to him before he entered their territory and agreed to provide food for his army, in return for more lenient treatment. As Alexander advanced into the Persian Empire, this was a deal that local governors were increasingly happy to agree to.

Conversely, removing or destroying all food and fodder in the path of an advancing army (a so-called scorched-earth policy) provided a way to use food defensively. An early example came during the Second Punic War between Rome and Carthage, during which Hannibal, the Carthaginian general, humiliated the Romans by rampaging around Italy with his army for several years. In an effort to stop him, a proclamation was issued that "all the population settled in the districts through which Hannibal was likely to march should abandon their farms, after first burning their houses and destroying their produce, so that he might not have any supplies to fall back upon." This ploy failed, but on other occasions in history it was highly effective. Another defensive strategy was to deny the enemy access to food-processing equipment. In order to delay the advance of Spanish troops in 1636, French generals were instructed to "send out before them seven or eight companies of cavalry in a number of places, with workers to break all the ovens and mills in an area stretching from their own fronts to as close as possible to the enemy." Without ovens and mills, seized grain could not be turned into bread, and soldiers would have to make camp for a couple of days to set up portable ovens.

All these food-related constraints on the waging of war persisted throughout most of human history, despite the emergence of new technologies such as firearms. But over time the supply systems used by armies invariably became more elaborate. In particular, warfare in eighteenth-century Europe became increasingly formalized, and armies came to rely less on requisitions and foraging, which they regarded as old-fashioned and uncivilized, and more on supplies amassed in depots and delivered by wagon trains. Professional soldiers expected to be fed and paid while on campaign; they did not expect to have to forage for food. The resulting need to build up supplies beforehand meant that campaigns had to be planned long in advance. With armies tethered to their supply depots, lightning strikes or long marches were out of the question. One historian has likened wars of this period to "the jousting of turtles."

The American Revolutionary War of 1775–1783 provides a microcosm of how logistical considerations could still be crucial in determining the outcome of a conflict, centuries after Alexander and Hannibal. In theory, the British should easily have been able to put down the rebellion among their American colonists. Britain was the greatest military and naval power of its day, presiding over a vast empire. In practice, however, supplying an army of tens of thousands of men operating some three thousand miles away posed enormous difficulties. Britain's 35,000 soldiers required 37 tons of food a day among them (a pound of beef each, plus some peas, bread, and rum); their 4,000 horses needed a further 57 tons.

To start with, the British commanders expected their soldiers' reliance on supplies delivered across the Atlantic by ship to be temporary. They hoped that American loyalists would rally to their cause, allowing the army to draw food and fodder from the country in loyalist areas. But this proved to be impractical, both because of the quantities required and because requisitioning food alienated the loyalists on whose support the British strategy depended. Many of the British troops, accustomed to Europe's more formal style of warfare, lacked experience in foraging and felt that it was beneath them. The British troops found themselves penned up near ports for security, dependent on supplies brought in by sea and unable to move very far inland. Attempts to enlarge the area under control provided a larger area in which to forage, but it caused resentment among the colonists, who refused to continue food production or mounted guerrilla resistance. Foraging expeditions sent beyond the British lines required covering forces of hundreds of troops. A small group of rebels could harass a much larger foraging party, picking off men using ambushes and snipers. The British lost as many men in such skirmishes as they did in larger pitched battles.

Unwilling to venture inland, where their movements would end up being determined by the needs of supply rather than military strategy, the British concluded that they would need to build up a reserve of at least six months' worth of food (and ideally a year's worth) before mounting a major offensive, a condition that was met only twice over the course of the eight-year war. The shortage of supplies also meant that the British were unable to press their advantage when the opportunity arose, repeatedly giving their opponents the chance to regroup. The British failed to strike a decisive blow in the early years of the conflict, and after other European powers entered the war on America's side it became clear that Britain could not win.

The American forces also suffered from supply problems of their own, but they had the advantage of being on familiar territory, and could draw manpower and supplies from the country in a way the British could not. As George Washington, the commander in chief of the American forces, remarked shortly afterward, "It will not be believed that such a force as Great Britain has employed for eight years in this Country could be baffled in their plan . . . by numbers infinitely less, composed of men sometimes half starved; always in rags, without pay, and experiencing, at times, every species of distress which human nature is capable of undergoing." The British failure to provide adequate food supplies to its troops was not the only cause of its defeat, and of America's subsequent independence. But it was a very significant one. Logistical considerations alone do not determine the outcome of military conflicts, but unless an army is properly fed, it cannot get to the battlefield in the first place. Adequate food is a necessary, though not sufficient, condition for victory. As the Duke of Wellington put it: "To gain your [objectives] you must feed."

"An Army Marches on Its Stomach"

In the early hours of October 5, 1795, a promising young artillery officer named Napoleon Bonaparte was put in charge of the forces defending the French government, known as the National Convention. It had been elected in 1792, in the wake of the French Revolution that had overthrown the monarchy, but there were still large numbers of royalist sympathizers in the country. An army of thirty thousand royalists was now advancing on the Tuileries Palace in Paris, where the convention's members had taken refuge. Napoleon immediately sent a cavalry officer to fetch forty cannons and their crews, and by dawn he had positioned them carefully in the streets around the palace and had them loaded with grapeshot. His defending forces were outnumbered six to one, and at one point Napoleon had his horse shot out from under him as he directed his men. When the royalist columns launched their main attack, the defending troops managed to channel them toward the main bank of guns, positioned in front of a church. Napoleon gave the order to fire, and the cannons cut down the royalist troops with devastating effectiveness, causing the survivors to turn and flee. "As usual I did not receive a scratch. I could not be happier," Napoleon wrote to his brother Joseph afterward. It was to prove a turning point in his career.

A few days later General Paul Barras, who had delegated the defense of the government to Napoleon, appeared with him and other officers before the convention's members, who wanted to express their thanks. Without warning one of the politicians climbed up to the dais to speak, and instead of thanking Barras, he declared that the hero of the hour had in fact been "General Bonaparte, who had only that morning in which to station his cannon so cleverly." Napoleon instantly became a celebrity, applauded whenever he appeared in public, and he was rewarded soon afterward with the command of the French forces in Italy. In the months that followed he waged a rapid and brutal campaign against the Austrians, bringing most of northern Italy under French control. Napoleon even dictated the terms of the peace with the Austrians, despite lacking the formal authority to do so. He became a national hero in France and used his success on the battlefield to win political influence in Paris, paving the way for his seizure of power in 1799. After his Italian campaign one French general even described him as "a new Alexander the Great."

This was in fact quite an accurate description, because one of the main things that distinguished Napoleon from other generals of his day, and shaped the course of his career, was the readoption of Alexander's minimalist approach to logistics. As a French general, the Comte de Guibert, had pointed out in the 1770s, armies of the period had become terribly reliant on their cumbersome supply systems and depots, or magazines. He suggested that they ought to be more mobile, travel light, and live off the country. Guibert also observed that relying on standing armies of professional soldiers meant that most ordinary people were untrained in the use of arms. He predicted that the first European nation to develop a "vigorous citizen soldiery" would triumph over the others. In the event his ideas prevailed, but not because of a deliberate program of military reform. Instead the French Revolution in 1789 resulted in the collapse of the old supply system and forced French soldiers fighting in the wars that followed to fend for themselves.

Reliance on living off the land began as a necessity, but the French army soon developed it into an organized system of requisitioning and amassing food, fodder, and other supplies. As Napoleon himself explained to one of his generals: "It is up to the commanding generals to obtain their provender from the territories through which they pass." Individual companies would send out eight or ten men under the command of a corporal or sergeant, for as little as a day or as a long as a week. These foraging parties would spread out behind the vanguard of the advancing army and requisition food from nearby villages and farms, sometimes paying for it with gold, but more often with an assignat, or receipt, that could in theory be presented for reimbursement once hostilities had ended. (The expression "as worthless as an assignat" indicates how rarely this happened in practice.) The foragers would then return to their companies to distribute what they had collected, with the food often being made into a stew or soup. This resulted in much less waste than the disorganized pillaging of the past, and French soldiers quickly became experts at finding hidden stores and evaluating how much food was available in a given area. "The inhabitants had buried everything underground in the forests and in their houses," observed one French soldier of the time. "After much searching we discovered their hiding places. By sounding with the butt ends of our guns we found provisions of all sorts."

All this made French armies extremely agile; they needed around one eighth of the number of wagons used by other armies of the time, and were capable of marching fifty miles per day, at least for a day or two. Greater mobility dovetailed neatly with Napoleon's military strategy, encapsulated in the maxim "divide for foraging, concentrate for fighting." His preferred approach was to split up his forces, spreading them out over a wide front to ensure each fast-moving corps had its own area in which to forage, and then suddenly concentrating his troops to force the enemy into a decisive battle. The result was a stunning series of French victories that gave the French army under Napoleon a fearsome reputation.

Napoleon did not do away with traditional supply systems altogether, however. When preparing for a campaign he would have large depots prepared within friendly territory, to provide supplies for his troops as they crossed the border. Soldiers carried a few days' worth of supplies, usually in the form of bread or biscuits, for use when foraging could not provide enough food, or when the enemy was nearby and the French forces were concentrated. As Napoleon himself observed, "the method of feeding on the march becomes impracticable when many troops are concentrated."

The best example of how all this worked came in the autumn of 1805, in the campaign that culminated in the battle of Austerlitz. Having amassed a large army in northern France with the intention of invading Britain, Napoleon instead found himself threatened by Britain's allies, Austria and Russia, and ordered his troops to head east through France. Mayors of towns along the way, two or three days apart, were asked to provide provisions for distribution to the soldiers as they passed through. Meanwhile, Napoleon ordered 500,000 biscuit rations to be prepared in cities along the Rhine. A month after being mobilized, Napoleon's 200,000 troops crossed the Rhine, spread over a front more than one hundred miles across. Each corps was instructed to live off the country to its left, requisitioning supplies from the local people and issuing receipts in the standard French way. Records show just how much food the French were able to extract, even from small towns. The German city of Heilbronn, with a population of around 15,000, produced 85,000 bread rations, 11 tons of salt, 3,600 bushels of hay, 6,000 sacks of oats, 5,000 pints of wine, 800 bushels of straw, and 100 wagons to carry what was not immediately consumed. The city of Hall, with only 8,000 inhabitants, produced 60,000 bread rations, 70 oxen, 4,000 pints of wine and 100,000 bundles of hay and straw. It helped that the French campaign occurred at harvest time, which meant more supplies were available than at any other time of year. Preparing and delivering supplies for such a large army using depots and wagon trains alone, in the eighteenth-century style, would have taken months to organize and would have prevented the army from moving so quickly.

Napoleon's aim was to defeat the Austrian army in the Danube region before the Russians arrived to reinforce it. He accomplished this with the celebrated "Ulm maneuver": Cavalry attacking from the west distracted the Austrian army while the main French force swiftly marched around it, encircling the Austrians and forcing them to surrender. Having taken care of the Austrians, Napoleon then set off in pursuit of the Russian army. This meant traveling through wooded country where there was little food to be had, so Napoleon issued his men with eight days' rations in bread and biscuits, gathered from the region around Ulm. This sustained his army until it reached richer territory to the east, where it could once again make requisitions; several Austrian depots were also captured. Once Vienna, the Austrian capital, had been taken it could be used as a supply depot, providing vast amounts of food and fodder: 33 tons of bread, 11 tons of meat, 90 tons of oats, 125 tons of hay, and 375 buckets of wine were requisitioned on one day alone. The army was given three days to recuperate before heading north in pursuit of the Russians, now joined by the remaining Austrian forces. The two armies eventually took up positions facing each other near the city of Austerlitz (modern Slavkov, in the Czech Republic), and Napoleon's subsequent victory is widely regarded as the greatest of his career. Napoleon had advanced deep inside enemy territory and had prevailed, humiliating the Austrian Empire. His army's unrivaled speed and mobility, made possible by its ability to break free when necessary from traditional supply systems, played a decisive role in his triumph. As Napoleon himself is said to have observed, "An army marches on its stomach."

Having underpinned his greatest victory, however, food also contributed to Napoleon's greatest blunder: his invasion of Russia in 1812. As he began planning the campaign in 1811, it is clear that

Napoleon did not expect his troops to be able to live off the land once they crossed into Russia. He ordered large supply depots to be established in Prussia and expanded the French military train with the addition of thousands of new wagons. And he proposed switching from four-horse to six-horse wagons, with 50 percent greater capacity, to reduce the number of wagons needed to carry a given amount of food. By March 1812 enough supplies had been gathered in the city of Danzig to supply four hundred thousand men and fifty thousand horses for seven weeks, and more supplies were being gathered along the Polish border. Napoleon hoped to mount a swift, decisive campaign, engaging the Russian army near the border and defeating it swiftly. He did not expect his army to have to venture very far into Russia, or to have to depend on foraging for food.

Napoleon's army of 450,000 crossed into Russian territory in late June 1812, carrying twenty-four days' worth of supplies: The men carried four days of rations in their packs, and the rest was in wagons. The problems began almost immediately. Heavy rain turned the poor local roads, little more than dirt tracks, into muddy swamps. The heavy wagons quickly became bogged down, horses broke their legs, and men lost their boots. The infantry moved more quickly, some units advancing seventy miles in two days, but they were then cut off from their supplies. Once the soldiers had consumed the rations they were carrying with them, they had to resort to living off the land. But the countryside was poor, and the army included many inexperienced recruits who were unfamiliar with the efficient French system of foraging. Discipline broke down and instead of careful distribution of supplies there was indiscriminate plunder. The few villages and farms along the route were soon exhausted of food, there was not enough grass to provide fodder for the French horses, and the crops in the fields were not ripe enough to harvest. "The advance guard lived quite well, but the rest of the army was dying of hunger," a French general later recalled.

The Russians retreated as the French advanced, abandoning their positions and falling back toward Moscow. Napoleon expected the richer country around Smolensk and Moscow to be able to provide food for his army, so he pressed on. But the Russians were stripping the countryside and destroying supplies as they retreated. The French army began to disintegrate as the men, weakened by hunger, fell prey to disease. A Russian general observed: "The roads were strewn with the carcasses of horses, and swarming with sick and stragglers. All French prisoners were carefully questioned as to the matter of subsistence; it was ascertained that already, in the neighborhood of Vitebsk, the horses were obtaining only green forage, and the men, instead of bread, only flour, which they were obliged to cook into soup." By the end of July, a mere five weeks after the start of the campaign, the French army had lost 130,000 men and 80,000 horses, and had yet to bring the enemy to battle. In August an indecisive battle was fought at Smolensk, which fell to the French, but only after the Russians had destroyed all supplies of food in the city. A far bloodier battle at Borodino ended with a Russian retreat, leaving the road open to the capital.

By denying Napoleon a decisive victory Prince Mikhail Illarionovich Kutuzov, the Russian commander, forced him to move even deeper into Russia, worsening his supply problems which, the Russians knew, posed the greatest threat to Napoleon's soldiers. Upon his arrival in Moscow with one hundred thousand remaining troops, Napoleon expected to be met by the city elders, but instead he found the city abandoned, with no civil administration to organize the collection of supplies for the army. Fires were already burning when the French arrived, and they turned into a huge conflagration, destroying three quarters of the city and many of its stores of food. (As well as setting fires, the retreating inhabitants of Moscow had also destroyed all the fire-fighting equipment.) The capture of the Russian capital proved to be a worthless victory: Napoleon had expected the Russians to capitulate and sue for peace, but he soon realized that they had no intention of doing so. The longer the French remained in the city, the more vulnerable they would become. A month after its arrival, the army began its retreat westward, accompanied by thousands of wagons

Napoleon's retreat from Moscow.

loaded with loot. But treasure cannot be eaten, and the shortage of food prompted infighting and further desertions.

Discipline collapsed and the army dissolved into a disorderly, ragtag horde thinking only of its own survival, weakened by hunger and illness and reduced to eating dogs and horses. Stragglers were set upon by Cossacks and tortured to death by local peasants. Abandoned wagons and cannons littered the roads. "If I met anyone in the woods with a loaf of bread I would force him to give me half— no, I would kill him and take it all," wrote one French soldier. The winter set in later than usual, in early November, toppling horses on icy roads and freezing men to death as they camped out at night. It is sometimes claimed that the Russian winter was responsible for Napoleon's defeat, but it merely hastened the destruction of his army, a process that was already well advanced. Only around 25,000 of Napoleon's main force of 450,000 troops eventually withdrew from Russia in December 1812. Napoleon had been defeated, and the myth of his invincibility had been shattered. His command of logistics had helped to make him the ruler of most of Europe, but it failed him in Russia and marked the beginning of his decline.

The Invention of Canned Food

In 1795, in an effort to improve the diets of soldiers and sailors during military campaigns, the French government offered a prize to anyone who could develop a new way to preserve food. The rules stipulated that the resulting food should be cheap to produce, easy to transport, and better tasting and more nutritious than food preserved using existing techniques. Salting, drying, and smoking had all been used to preserve foodstuffs for centuries, but all of them affected the taste of food and failed to preserve many of its nutrients. Experiments to find better ways to preserve food had been going on since the seventeenth century, when scientists had begun to take an interest in the process of decomposition and, by extension, how it could be prevented.

Robert Boyle, an Irish scientist known as the "Father of Chemistry," developed a vacuum pump and made many discoveries with it, showing for example that the sound of a ringing bell inside a sealed jar diminished in volume as the air was pumped out. Boyle also speculated that the decomposition of food was dependent on the presence of air, and he tried preserving food by storing it in evacuated jars. But he eventually concluded that contact with air was not the sole cause of decomposition. Denis Papin, a French physicist, extended Boyle's work by sealing food in evacuated bottles and then heating them. This seemed to work much better, though the food still spoiled sometimes. From time to time Papin would present his preserved food to other scientists at meetings of the Royal Society in London. In 1687 they reported that he had preserved "great quantities" of fruit: "He shuts up the Fruits in Glass Vessels exhausted of the Air, and then puts the Vessel thus exhausted in hot Water, and lets it stand there for some while; and that is enough to keep the Fruit from the Fermentation, which would otherwise undoubtedly happen."

At the time the mechanism of decomposition was not understood, though many people subscribed to the theory of "spontaneous generation," an idea going back to the Greeks which held that maggots were somehow generated from decomposing meat, mice from rotting piles of grain, and so on. Despite the experimental work of Boyle, Papin, and others, the problem of food preservation remained unsolved. The various preservation techniques developed during the seventeenth and eighteenth centuries were both expensive and unreliable. Nobody managed to improve upon the traditional military rations of salted meat and dry biscuits, which explains the conditions attached to the French prize in 1795.

The man who eventually claimed the prize was not a scientist but a cook. Nicolas Appert was born in Châlons-sur-Marne, on the edge of France's Champagne region, in 1749. His father was a hotelier, and he became an accomplished chef, serving in the kitchens of various noblemen before setting up as a confectioner in Paris in 1781. In this line of work he was necessarily aware of the use of sugar to preserve fruit, and he wondered whether it could be used to preserve other foods. As his interest in food preservation grew he began to experiment with storing food in sealed champagne bottles. In 1795 he moved to the village of Ivry-sur-Seine, where he began to offer preserved foods for sale, and in 1804 he set up a small factory. By this time some of his preserved food had been tested by the French navy, which was impressed by its quality. "The broth in bottles was good, the broth with boiled beef in another bottle very good as well, but a little weak; the beef itself was very edible," its report concluded. "The beans and green peas, both with and without meat, have all the freshness and flavor of freshly picked vegetables."

Appert later described his method as follows. "First, enclose the substances you wish to preserve in bottles or jars; second, close the openings of your vessels with the greatest care, for success depends principally on the seal; third, submit the substances, thus enclosed, to the action of boiling water in a bain-marie . . . fourth, remove the bottles from the bain-marie at the appropriate time." He listed the times necessary to boil different foods, typically several hours. Appert was not familiar with the earlier work of Boyle, Papin, and others; he had devised his method solely by experiment and had no idea why it worked. It was not until the 1860s that Louis Pasteur, a French chemist, finally determined that decomposition was caused by microbes that could be killed by applying heat. That is why Papin's technique, which involved heating, had worked; but most of the time he had not heated his food samples enough to kill off the microbes. Appert's long process of trial and error had revealed that heat had to be applied for several hours in most cases, and that some foods needed to be heated for longer than others. "The application of fire in a manner variously adapted to various substances, after having with the utmost care and as completely as possible, deprived them of all contact with the air, effects a perfect preservation of those same productions, with all their natural qualities," he concluded.

Word of Appert's products spread and they went on sale as luxury items in Paris; his factory was soon employing forty women to prepare food, put it into bottles wrapped in cloth bags in case of breakage, and then boil the bottles in vast cauldrons. Meanwhile military trials continued, and in 1809 Appert was invited to demonstrate his method to a government committee. He prepared several bottles of food as the officials watched, and a month later they returned to taste the contents, which were found to be in excellent condition. Appert was duly awarded the prize of twelve thousand francs on the condition that he publish the details of his method in full, so that it could be widely adopted throughout France. Appert agreed, and his book, *The Art of Preserving All Kinds of Animal and Vegetable Substances for Several Years*, appeared in 1810. In accepting the government prize, Appert agreed not to patent his method in France.

Within three months of his book's publication, however, a businessman in London, Peter Durand, had been granted an English patent for a preservation technique that was essentially identical to Appert's. Durand sold the patent to an engineer named Bryan Donkin for one thousand pounds, and Donkin set up a company in conjunction with two partners involved in an iron works. Instead of preserving food in bottles, Donkin's firm used canisters made of tin-coated iron, known today as tin cans. Durand admitted that the technique was "an invention communicated to me by a certain foreigner," and it has long been assumed that he simply stole Appert's idea. More recent research has indicated, however, that Durand may in fact have been acting on Appert's behalf in England, and arranged to patent his invention and sell the rights. Appert even visited London in 1814, probably to collect his share of the proceeds from Durand. By this time the Royal Navy had tested the new canned food, and samples had even been presented to the royal family. But Appert came away from London empty-handed. His English partners appear to have cut him out of the deal; he could hardly expose them, since he had been trying to profit by selling his invention to an enemy nation.

Appert concentrated instead on refining his process and supplying the French army and navy. He embraced the use of tin cans for military supplies, but he continued to sell food in glass bottles to civilian customers. One French explorer, who took Appert's canned food on a three-year voyage, declared that the invention had "completely resolved the problem of feeding sailors." Canned food had obvious military advantages. It allowed large numbers of rations to be prepared and stockpiled in advance, stored for long periods, and transported to combatants without the risk of spoiling. Canning could smooth over seasonal variations in the availability of food, allowing campaigns to continue through the winter. The new technology was adopted very quickly: Some of the soldiers on the battlefield at Waterloo in 1815, the scene of Napoleon's final defeat, carried canned rations. Canned meat fed English and French troops in the Crimean War, and tinned meat, milk, and vegetables were supplied to Union soldiers in the American Civil War. Soldiers have carried canned rations of various kinds ever since. The early cans had to be opened with a hammer and chisel, or using a bayonet. The first can openers appeared only in the 1860s, when canned food started to become popular among civilians.

As far as the civilian population was concerned, canned food was still a novelty or luxury item. At the Great Exhibition in London in 1851, the company founded by Bryan Donkin some four decades earlier displayed "canisters of preserved fresh beef, mutton and veal; of fresh milk, cream and custards; of fresh carrots, green peas, turnips, beetroots, stewed mushrooms and other vegetables; of fresh salmon, codfish, oysters, haddock and other fish . . . Preserved hams for use in India, China, etc . . . all preserved by the same process . . . The whole preserved so as to keep in any climate, and for an unlimited length of time." Expensive preserved foods, including truffles and artichokes, were also exhibited by Appert's company, now run by his nephew.

But canned foods did not remain luxuries for much longer. Strong military demand prompted inventors to devise new machinery to automate the process of sealing cans, and it was found that adding calcium chloride to the water in which they were treated raised its boiling point and reduced the boiling time required. As volumes increased and prices fell, canned food became more widely affordable. In America, the production of canned food went from five million cans a year to thirty million between 1860 and 1870; in Britain, an outbreak of cattle disease in the 1860s prompted people to turn to canned meat from Australia and South America. Appert died in 1841 at the age of ninety-one, but his method of preserving food, heat-treated in a sealed container, and inspired by the supply difficulties of the French Revolutionary army, is still in use today.

"Forage Liberally"

Canned food was one of two inventions that transformed military logistics during the nineteenth century. The second was mechanized transport, in the form of the railway and the steam locomotive, which could move troops, food, and ammunition from one place to another at unprecedented speed. This meant an army could be resupplied easily—provided it did not stray far from a railway line. The impact of this new development became apparent during the American Civil War, a transitional conflict in which old and new approaches to logistics appeared side by side.

When the war began in 1861 there were thirty thousand miles of railway track in America, more than in the rest of the world combined.

More than two thirds of this track was in the more industrialized northern states of the Union, giving the North a clear advantage in supplying its troops. The Union's strategy was to blockade the break-away southern states of the Confederacy in an effort to cause food shortages and economic collapse. A blockade of southern ports was imposed in 1861, and the Union then set about seizing control of the Mississippi River and disrupting the southern rail networks, in order to hinder the distribution of food and supplies. Between 1861 and 1863 the prices of some basic foodstuffs increased sevenfold, causing riots in several southern cities in which angry mobs attacked grocery stores and warehouses. With many basic foodstuffs unavailable, various ingenious substitutes were devised, and both soldiers and civilians resorted to eating anything they could lay their hands on. One Confederate soldier wrote to his wife in 1862: "We have lived some days on raw, baked and roasted apples, sometimes on green corn and sometimes nothing."

By the time Ulysses S. Grant was put in charge of all Union forces in 1864, the Confederacy had suffered several significant defeats and the blockade was causing severe food shortages. Grant devised a two-pronged plan to end the war: a large Union force would take on the main Confederate army commanded by Robert E. Lee, and smaller Union forces would meanwhile undermine morale in the South by attacking agricultural regions and cutting railway links to further aggravate the shortages. Accordingly, Union forces attacked the agriculturally rich Shenandoah Valley, an important source of supplies to the Confederate forces, and conducted a scorched-earth campaign, destroying crops, barns, and mills. But it is the campaigns undertaken by William Sherman in Georgia and the Carolinas that highlight how much the field of military logistics had changed—and how much it had not.

Sherman was under instructions from Grant "to get into the interior of the enemy's country as far as you can, inflicting all the damage you can against their war resources." After stockpiling supplies in Nashville, Tennessee, Sherman began the march south toward Atlanta, Georgia, in May 1864, following the line of the railway so that food, fodder, and ammunition could be delivered to his army by train. Special teams of engineers repaired the track as the retreating Confederate army attempted to sabotage it. As he moved south through Georgia, Sherman established new bases in Marietta and Allatoona, supplied by railway from Nashville which lay farther up the the line. In July he informed Grant that "we have been wonderfully supplied in provisions and ammunition; not a day has a regiment been without bread and essentials. Forage has been the hardest, and we have cleaned the country in a breadth of thirty miles of grain and grass. Now the corn is getting a size which makes a good fodder, and the railroad has brought us grain to the extent of four pounds per animal per day."

The age-old difficulty of finding enough fodder for animals remained, but when it came to food and ammunition, Sherman's army was exploiting a state-of-the-art logistics system. Delivering supplies from the rear by rail was a far faster and more reliable alternative to the supply wagons, shuttling between the army and its nearest supply depot, that soldiers had depended on for centuries. Sherman's men only needed to carry a few days' worth of supplies to sustain them between rail deliveries. The rail link also meant that ammunition could be delivered in large quantities; Sherman's army was consuming hundreds of thousands of rounds per day as it fought its way toward Atlanta. Military logistics was starting to shift toward providing supplies for machines, rather than for men and animals.

Having arrived in the vicinity of Atlanta, Sherman concentrated his efforts on seizing control of the converging railway tracks that connected the city to the rest of the Confederacy. He was prepared to mount a long siege, since he was confident of being able to supply his troops by rail from the north. But as things turned out, he captured the railway lines within a few weeks and the Confederate army abandoned Atlanta. Sherman occupied the city and planned the next stage in his campaign, known as the "March to the Sea." By contrast with the modernity of his advance on Atlanta, this was to be a rather more old-fashioned stratagem. The plan was to cut loose from the formal supply system and march three hundred miles through Georgia to Savannah, on the Atlantic coast, destroying as much agricultural and economic infrastructure as possible along the way. The army would then head north through the Carolinas to prevent reinforcements reaching Lee's army, which was besieged at Petersburg, Virginia. Sherman's troops would carry some rations with them, but they would live off the land as much as possible, destroying what they could not eat. This, one of the last and most effective campaigns of the Civil War, is a striking (some would say infamous) example of the use of food as a weapon. Sherman issued a special field order:

> The army will forage liberally on the country during the march. To this end, each brigade commander will organize a good and sufficient foraging party, under the command of one or more discreet officers, who will gather, near the route traveled, corn or forage of any kind, meat of any kind, vegetables, corn-meal, or whatever is needed by the command, aiming at all times to keep in the wagons at least ten days' provisions for the command and three days' forage. Soldiers must not enter the dwellings of the inhabitants, or commit any trespass, but during a halt or a camp they may be permitted to gather turnips, potatoes, and other vegetables, and to drive in stock of their camp. To regular foraging parties must be instructed the gathering of provisions and forage at any distance from the road traveled.

The march began in November, just after the harvest, so the barns were full of grain, fodder, and cotton. Each brigade sent out a foraging party of "bummers" who would set out on foot and return with wagons of food, driving cattle in front of them. Sherman's troops fanned out and devastated the country, helping themselves to fresh mutton, bacon, turkeys, chickens, cornmeal, and sweet potatoes, among other things. As well as taking the supplies they needed to subsist, the Union soldiers killed pigs, sheep, and poultry and burned and looted many houses, despite their orders to the contrary. They were instructed to destroy mills, barns, and cotton gins only if they encountered any resistance. Sherman recalled in his memoirs that the foraging became general plunder, and was not limited to formal foraging parties as he had ordered: "A soldier passed me with a ham on his musket, a jug of sorghum—molasses—under his arm and a big piece of honey in his hand, from which he was eating and, catching my eye he remarked in a low voice to a comrade, 'Forage liberally on the country.' " Sherman claimed to disapprove of such lawlessness, but it was entirely in keeping with his boast to Grant that he would "make Georgia howl."

As well as plundering and destroying farms and mills, the Union solders tore up railway tracks whenever they encountered them and devised elaborate tricks to ensure that they could not be repaired, such as heating and warping the rails and wrapping them around the trunks of trees. This inflicted hardship not just on the people of Georgia, but also on the Confederate armies who relied on their produce, since supplies could no longer be delivered by rail. Sherman's army also damaged the southern economy by liberating black slaves, thousands of whom followed the army as it marched.

Sherman's march spread fear and confusion, not least because his destination was unclear. By the time it became clear that he was heading for Savannah, the Confederate armies were unable to concentrate their forces to stop him. The Union soldiers met little resistance, and attempts by the authorities to organize a scorched-earth defense ("Remove your negroes, horses, cattle, and provisions from Sherman's army and burn what you cannot carry away") failed; morale had collapsed, and with it confidence in the government. On his arrival in Savannah, Sherman reported that "we have consumed the corn and fodder in the region of country thirty miles on either side of a line from Atlanta to Savannah as also the sweet potatoes, cattle, hogs, sheep and poultry, and have carried away more than 10,000 horses and mules as well as a countless number of their slaves. I estimate the damage done to the State of Georgia and its military resources at

$100,000,000; at least $20,000,000 of which has inured to our advantage and the remainder is simple waste and destruction."

More was to come. Sherman then continued his destructive march northward through the Carolinas in the spring of 1865, leaving a trail of destruction forty miles wide. "Sherman's campaign has produced bad effect on our people," conceded Jefferson Davis, the president of the Confederacy. Lee reported an "alarming frequency of desertions" from his Confederate army, chiefly due to the "insufficiency of food and non-payment of the troops." Lee realized his position was untenable and surrendered, and the rest of the Confederate forces soon followed, ending the war.

FOOD FOR MACHINES

The American Civil War encapsulated the shift from the Napoleonic era of warfare to the industrialized warfare of the twentieth century. As Sherman's men advanced through Georgia, living off the land as armies had done for thousands of years, the opposing armies of Grant and Lee were engaged in trench warfare around Petersburg, their zigzag fortifications prefiguring the elaborate ditches and tunnels that would scar the fields of France during the First World War. The emergence of trench warfare was a consequence of improvements in the range, power, and accuracy of firearms and artillery that were not matched by corresponding improvements in mobility. Armies had unprecedented firepower at their disposal—provided they did not move. For most of history, an army that stayed still risked starvation, unless it could be supplied by sea. But the advent of canned food and railways meant that soldiers could be fed all year round, and for as long as necessary, as they stayed put in their trenches.

Even so, for most of the First World War the new logistics coexisted with the old. Ammunition and food for the front were delivered by rail; but the only way to carry supplies over the last few miles from the railhead to the front line was by using horse-drawn wagons. Accordingly, enormous quantities of fodder also had to be sent by rail, and an ancient logistical constraint survived into the twentieth century: Fodder was the largest category of cargo unloaded at French ports for the British army during the war. The stalemate of trench warfare ended only with the development of the tank, which coupled greater firepower with mobility and heralded a new era of motorized warfare in which fuel and ammunition, to feed vehicles and weapons, displaced food for men and animals as the most important fuel of war.

This was vividly illustrated just a few years later during the Second World War, and on the North African front in particular, where the German general Erwin Rommel found himself hemmed in by logistical constraints—primarily that of fuel. The German and Italian troops in North Africa received supplies via the port of Tripoli. Rommel dreamed of defeating the British, based to his east in Egypt, and then choking off the Allies' supply of oil from the Middle East. But there was no suitable railway line along which he could advance to the east, so his supplies had to be carried across the desert in trucks. As the German troops advanced, convoys of trucks shuttled back and forth between Tripoli and the front, carrying fuel, ammunition, food, and water. Seizing a deep-water port along the coast would reduce the distance that supplies needed to be carried overland, so Rommel captured the Libyan port of Tobruk, near the border with Egypt. But the port's capacity was limited and approaching ships were sunk by the Allies in large numbers. Rommel's supply lines were so overextended that 30 to 50 percent of his fuel was being used to ferry fuel and other supplies to the front. The farther east he advanced, the more fuel was wasted in this way. When he retreated or was pushed back westward, his supply problems eased.

Rommel's attempt to defeat the Allies in North Africa failed. "The first essential condition for an army to be able to stand the strain of battle is an adequate stock of weapons, petrol, and ammunition," he eventually concluded. "In fact, the battle is fought and decided by the quartermasters before the shooting begins." In a previous era he would have mentioned food and fodder. But they were no longer the critical elements of military supply. Food's central role in military planning had come to an end. But by the middle of the twentieth century food was already taking on a new role: as an ideological weapon.

10

FOOD FIGHT

Food is a weapon.

How do you deal with mice in the Kremlin? Put up a sign saying "collective farm." Then half the mice will starve, and the other half will run away.

FOOD FROM THE SKY

The Cold War between the United States and the Soviet Union, an ideological struggle between capitalism and communism that overshadowed the second half of the twentieth century, began in earnest with a food fight over the city of Berlin. Germany had been divided at the end of the Second World War into four zones—those controlled by Britain, France, and the United States in the west, and a fourth zone controlled by the Soviet Union in the east. Its capital, Berlin, situated in the heart of the Soviet zone, had also been divided in four in this way. In early 1948, nearly three years after the end of the war, the British, French, and Americans agreed to unite their respective zones of Germany, and of Berlin, under a single administration in order to coordinate the reconstruction of the country. The Soviets were strongly opposed to the Western allies' plan, because Germany had emerged as a symbolic battleground on which, both sides agreed, the future political direction of Europe would be decided. The Western nations wanted to establish a democratic government in a reunified Germany, whereas Russia hoped to orchestrate the installation of a Communist regime. The disagreement between the two sides became focused on Berlin, an isolated Western toehold in the Soviet zone of eastern Germany. As Vyacheslav Molotov, the Soviet foreign minister, put it: "What happens to Berlin, happens to Germany; what happens to Germany, happens to Europe."

Determined to force the Western allies to abandon West Berlin, the Soviets started interfering with the delivery of food and other supplies to the city, interrupting road, rail, and barge traffic on various spurious pretexts. The Soviets calculated that the Western allies would prefer to give up the city rather than go to war to defend it. In April 1948 Lucius D. Clay, the highest ranking American military officer in Germany, told Omar Bradley, the U.S. Army chief of staff, that "if we mean that we are to hold Europe against communism, we must not budge. We can take humiliation and pressure short of war in Berlin without losing face. If we move, our position in Europe is threatened . . . and communism will run rampant. I believe the future of democracy requires us to stay here until forced out." In June, Clay underlined his position in a telegram sent to his superiors in Washington, D.C.: "We are convinced that our remaining in Berlin is essential to our prestige in Germany and in Europe," he declared. "Whether for good or bad, it has become a symbol of the American intent."

As Soviet interference with delivery of supplies to West Berlin continued, Clay proposed sending an infantry division to accompany a convoy of trucks through Soviet-controlled East Germany to the city as a show of strength. But his plan was regarded as too risky, since it might have sparked a firefight between American and Soviet troops that could have escalated into a broader conflict. When the introduction of a new currency in West Germany was announced on June 18, in effect formalizing the economic separation of East and West Germany, the Soviets expressed their displeasure by blocking freight access to West Berlin by road, rail, and barge. By the evening of June 24 all land and water access to West Berlin had been completely sealed off. Colonel Frank Howley, the U.S. commandant in Berlin, went on the radio to reassure the inhabitants of the city. "We are not getting out of Berlin, we are going to stay," he said. "I don't know the answer to the present problem—not yet—but this much I do know: The American people will not stand by and allow the German people to starve."

He was speaking unofficially, because the allies had not yet decided how to respond. But they had to do something: The city had only enough food for thirty-six days, and enough coal for forty-five days. Clay once again proposed his plan for an armed road convoy, and was again overruled. General Brian Robertson, the British commander in Germany, said that his government would not approve such a move either. But he suggested an alternative way to break the blockade: supplying West Berlin by air.

On the face of it, this was a preposterous idea. Supplying the two million people in West Berlin, it was calculated, would mean delivering some fifteen hundred tons of food and a further two thousand tons of coal and fuel every day, at a bare minimum. (Ideally, some 13,500 tons a day would be needed, but this was a minimum figure for the summer months.) The only aircraft available were Douglas C-47s, capable of carrying about three tons each. Even with the help of smaller British transports, it was hard to see how it would be possible to deliver the necessary volume of supplies. The airlift idea was, however, the only alternative to making a politically unacceptable climbdown and abandoning the city. It also had the advantage that, unlike the land-based access routes through East Germany to West Berlin, the status of which was legally unclear, the right to use air corridors to and from Berlin had been agreed in writing with the Soviet Union in November 1945. A small amount of supplies had in fact already been delivered by aircraft in April 1948, after the Soviets had begun interfering with rail freight.

So Clay ordered the airlift to begin. He assumed that he would be able to get hold of more planes fairly quickly, and that the airlift would only have to operate for a few weeks while a diplomatic solution to the crisis was agreed. The first aircraft, carrying supplies from airfields in West Germany, arrived in West Berlin on June 26. With the backing of President Harry Truman, who gave his formal support to the airlift despite objections from some of his advisers, the operation slowly scaled up, reaching twenty-five hundred tons a day by mid-July.

But diplomacy with the Soviet Union was getting nowhere. Tensions rose when America stationed B-29 bombers—the type of aircraft that had dropped atomic bombs on Japan in 1945—at airfields in Britain, within range of Moscow. The aircraft were not equipped with nuclear weapons, but the Soviets did not know this. After the airlift had been running for a month, however, the immediate threat of war seemed to have receded, and it had become clear that the airlift would have to operate for more than just a few weeks. The C-47s were replaced with larger C-54s, capable of carrying ten tons of cargo, and flights were soon operating every three minutes, twenty-four hours a day. General William H. Tunner, who was put in charge of the airlift in late July 1948, introduced new takeoff and landing rules to maximize capacity and minimize the risk of accidents. Teams of volunteers unloaded the aircraft in Berlin and competed to do so in the shortest possible time. The Americans called the mission "Operation Vittles"; to the British it was known as "Operation Plainfare." By October deliveries had reached five thousand tons per day.

The Soviets made various attempts to disrupt the airlift, harrassing the freight planes by buzzing them with their own aircraft, releasing barrage balloons that got in their way, causing radio interference, shining searchlights at incoming aircraft, and sometimes even firing into the air in their vicinity. But they never went so far as to shoot any of the planes down. The soldiers and airmen in Berlin, meanwhile, who had arrived in the city a few years earlier as an occupying enemy force, forged a close bond with the city's inhabitants, whose liberty they were now defending. Flying boats landing on a lake in central Berlin to deliver salt, which was too corrosive to be carried in other aircraft, were met by Berliners who paddled out to present their pilots with bunches of flowers. And an American pilot, Gail Halvorsen, be-

came a hero to the children of Berlin after he began dropping chocolate bars, sweets, and chewing gum, attached to parachutes made
from handkerchiefs, out of the window of his aircraft whenever he
passed over the city. Soon other pilots were following his example,
and Halvorsen's unofficial venture won official approval. Over three
tons of sweets, both supplied by American manufacturers and do-
nated by American children, were dropped on Berlin. Highlighting
the link between American children and those in Berlin, as their respective countries took a stand together against communism, gave
the operation enormous propaganda value.

That the food being supplied to West Berlin was being used, in effect, as a weapon against the Communists was explicitly acknowledged on a poster produced in 1949 by Douglas, the maker of the
C-54 planes that were the mainstays of the airlift. It shows a girl
holding up a glass of milk, and hundreds more glasses floating down
from passing aircraft in the sky. Under the headline MILK . . . NEW
WEAPON OF DEMOCRACY, the poster explains: "In today's diplomatic
Battle for Berlin, hope for democracy is being kept alive for millions
in Western Europe by the U.S. Air Force. Flying Douglas aircraft almost exclusively, Yankee crews have poured over half a million tons
of supplies into Berlin since last June."

In the spring of 1949 General Tunner decided to stage a spectacular
"Easter Parade" to demonstrate how committed the Allies were to continuing the airlift for as long as necessary. Deliveries were exceeding six
thousand tons a day by March 1949, but Tunner set the ambitious target of delivering ten thousand tons on a single day: April 17, which was
Easter Sunday. Maintenance schedules were arranged so that the max-
imum number of aircraft would be available that day, and crews at
different airfields prepared to break their previous records. The ground
crews and pilots were determined to beat the ten-thousand-ton target,
and in the event a total of 12,940 tons were delivered. This vividly
demonstrated the potential capacity of the airlift operation and the
commitment of the people operating it. The publicity surrounding the
Easter Parade sent a clear signal to the Soviets and helped to bring

"Milk . . . new weapon of Democracy"
poster produced by Douglas during the Berlin airlift.

about a new round of negotiations, at which the Soviets finally agreed
to lift the blockade of West Berlin from May 12, 1949. Delivery flights
did not end immediately, but they gradually wound down over several
months, to ensure that the operation could be stepped up again if necessary. The last flight took place on September 30. The airlift had
operated for fifteen months, during which some 2.3 million tons of
supplies were delivered in more than 275,000 flights.

Subsequent negotiations failed to reach agreement on the future of
Germany or Berlin. The crisis spurred the formation of the North
Atlantic Treaty Organization (NATO), a military alliance of Western

powers, on April 4, 1949, thus setting the stage for the standoff between America and its allies on the one hand, and the Soviet Union and its allies on the other, in the following decades. The first battle of this Cold War had been fought not with bullets or bombs, but with milk, sweets, salt, and other foodstuffs and supplies. In the four decades that followed there was never a direct conflict between NATO and Soviet forces. Instead the conflict was waged indirectly: through wars between the two sides' client states, through propaganda, and with ideological weapons—including food.

STALIN'S FAMINE

The Soviet leader, Josef Stalin, was no stranger to the use of food as an ideological tool. After assuming power in 1924 he had launched a crash industrialization program with the aim of catching up with, and then surpassing, the Western industrialized nations. Food was central to his plan. At the time, the Soviet Union was a major exporter of grain, and the purchase of industrial machinery from foreign countries was to be funded by an increase in such exports. Small farms run by individual farmers and their families would be crunched together to form "collective" farms owned by the state. Bringing farming under state control in this way would, Stalin hoped, boost production. "In some three years' time, our country will have become one of the richest granaries, if not the richest, in the whole world," Stalin declared in 1929, as he unveiled his plans. This would provide extra grain to sell abroad, yielding more hard currency to fund the industrialization program. Stalin set a goal of doubling steel output and tripling iron production within five years. The success of his program would demonstrate the superiority of socialism, as farmers working together produced more food and as the Soviet Union rapidly industrialized.

In some respects this was an attempt to reproduce what had happened in western Europe, starting in Britain, where industrialization had been preceded by a surge in agricultural productivity. This had liberated laborers from the land and made them available as industrial workers, which is why Adam Smith had called industrial activity "the offspring of agriculture." But the Soviet approach was very different, because the state had played a very limited role in orchestrating Britain's industrialization; it had not been a deliberately planned outcome. Stalin's industrialization program, by contrast, was a state-organized effort that would be funded by squeezing as much as possible out of peasant farmers. "Collectivizing" the farms would mean that their produce belonged to the state and therefore could be more readily appropriated for export.

Unsurprisingly the farmers themselves were less than enthusiastic about this new policy. Collectivization, in practice, meant herding the farmers into communal accommodation and, in some cases, forcing them to renounce private property and destroy their possessions. The more productive (and hence wealthier) farmers were particularly reluctant to go along with this. In some cases they chose to burn their crops or slaughter their cattle rather than surrender them to the new collective farms. Stalin decreed that since all crops, cattle, and agricultural produce now belonged to the state, anyone who refused to hand it over or destroyed it was an enemy of the people or a saboteur, and deserved to be deported to the Soviet network of penal labor camps, which later came to be known as the Gulag.

Since the most productive farmers were most likely to object to collectivization, the impact on agricultural productivity was predictable. With their produce now belonging to the state, there was no incentive for farmers to maximize production. Drought, bad weather, and a lack of horses to work in the fields also meant that the harvests of 1931 and 1932 were poorer than usual. The result was that just as Stalin was demanding more agricultural goods to fund his industrialization program, the level of food production actually fell. But admitting that collectivization had made farms less productive was unthinkable to the Soviet leadership. Stalin insisted instead that there had been record harvests, but that some farmers were hiding their produce to avoid having to hand it over. This explanation justified the state's continuing procurements of large amounts of grain. But it meant that many farmers were left without enough to eat. And those who failed to meet their grain quotas or were suspected of hiding grain were punished by having other crops removed as "fines," so that they had even less food. Meanwhile the industrial workers in the cities had plenty to eat, and exports of grain doubled, giving the outside world the impression that Stalin's scheme was proceeding as planned.

On average, farmers ended up with one third less grain for their own consumption than they had had before collectivization. But in some areas the situation was much worse. In particular, in Ukraine, a rich agricultural region that traditionally produced large grain surpluses, the state set ambitious procurement quotas. When the expected bumper harvests failed to materialize, local officials were ordered to step up their searches for hidden stores of food. Stalin decreed that retaining so much as one ear of wheat from the state was punishable by death or ten years' imprisonment. One participant recalled: "I took part in this myself, scouring the countryside, searching for hidden grain, testing the earth with an iron rod for loose spots that might lead to hidden grain. With the others I emptied out the old folks' storage chests, stopping my ears to the children's crying and the women's wails. For I was convinced that I was accomplishing the transformation of the countryside." As people began to starve, soldiers were posted to guard the large stores of grain that had been amassed by the state. Vasily Grossman, a Soviet writer, recorded the plight of those starving in rural villages: "People had swollen faces and legs and stomachs . . . and now they ate anything at all. They caught mice, rats, sparrows, ants, earthworms. They ground up bones into flour, and did the same thing with leather and shoe soles; they cut up old skins and furs to make noodles of a kind and they cooked glue. And when the grass came up, they began to dig up the roots and ate the leaves and buds."

In a speech in November 1932, Stalin argued that the difficulties with grain collection were being caused by saboteurs and "class enemies." He regarded this as a challenge to the authority of the regime by farmers who were deliberately obstructing his collectivization scheme. "It would be stupid if Communists . . . did not answer this blow, by some collective farmers and collective farms, with a knock-out blow," he declared. But sending hundreds of thousands of farmers to the Gulag would be difficult and expensive. Letting them starve was much easier. In another speech in February 1933, Stalin approvingly quoted Lenin's dictum "He who does not work, neither shall he eat." An official report in March stated: "The slogan 'He who does not work, neither shall he eat' is adopted by rural organizations without any adjustment—let them perish." Stalin did not initially intend collectivization to lead to starvation, but if "idlers" who refused to go along with it starved, that was, he implied, their own fault for being too lazy to grow enough food to feed themselves.

In early 1933 a system of internal passports was introduced to prevent people fleeing to the cities from the starving villages in Ukraine and the North Caucasus. Stalin also sent in agents of the OGPU, the state security agency, to step up the collection of grain in Ukraine, which he felt the local authorities were pursuing with insufficient vigor. A Politburo memo had complained of the "shameful collapse of grain collection in the more remote regions of Ukraine" and called for officials to "break up the sabotage of grain collection" and "eliminate the passivity and complacency toward the saboteurs." And a report sent to Stalin in March 1933 by Stanislav Kosior, who was in charge of the collectivization program in Ukraine, noted that the famine had not yet taught the peasants enough of a lesson. "The unsatisfactory preparation for sowing in the worst affected regions shows that the hunger has not yet taught many collective farmers good sense," Kosior declared.

Malcolm Muggeridge, a British journalist who visited Ukraine in May 1933, reported that officials "had gone over the country like a swarm of locusts and taken away everything edible; they had shot and exiled thousands of peasants, sometimes whole villages; they had reduced some of the most fertile land in the world to a melancholy desert." But his report was ridiculed by other journalists who had been taken on stage-managed visits to model communes and who insisted there was no famine. Yet in the Ukrainian capital of Kiev the Italian consul reported "a growing commerce in human meat," and the authorities were putting up posters saying EATING DEAD CHILDREN IS BARBARISM. At the same time, grain exports were increased in order to maintain the pretense that there was no problem, and that agriculture was booming under the Soviet regime. When some foreign aid organizations offered food aid, it was refused.

The political nature of the famine was most starkly outlined by Comrade Hatayevich, a senior official in the Ukraine, who explained in 1933 that "a ruthless struggle is going on between the peasantry and our regime. It's a struggle to the death. This year was a test of our strength and their endurance. It took a famine to show them who is master here. It has cost millions of lives, but the collective farm system is here to stay. We've won the war." It was a war waged by the regime against its own people, using food as a weapon. The famine ended in 1934 when Stalin scaled back the state procurements of grain and conceded that households should be allowed a small plot of land on which to grow vegetables and keep a cow, a pig, and up to ten sheep. These private plots, rather than collective farms, provided most of the country's food for the next fifty years.

Some seven to eight million people had died of starvation, the victims of Stalin's desire to maintain grain exports at all costs, both to convince the world of the superiority of communism and to fund Soviet industrialization. The famine's greatest impact was in Ukraine, where the millions of dead are now widely considered to have been the victims of genocide. One eyewitness, Fedor Belov, called the famine "the most terrible and destructive that the Ukrainian people have ever experienced. The peasants ate dogs, horses, rotten potatoes, the bark of trees, grass—anything they could find. Incidents of cannibalism were not uncommon. The people were like wild beasts, ready to devour one another. And no matter what they did, they went on dying, dying, dying. They died singly and in families. They died everywhere—in yards, on streetcars, and on trains. There was no one to bury these victims of the Stalinist famine. A man is capable of forgetting a great deal, but these terrible scenes of starvation will be forgotten by no one who saw them."

The Worst Famine in History

After the Communists, led by Mao Zedong, seized power in China in 1949, they were very keen to follow the Soviet model of collectivization, which had supposedly been such a success in increasing food production and underwriting industrialization. Leaflets, pamphlets, and propaganda films distributed in China lauded the Soviet triumph. As one Chinese woman later recalled: "We heard a lot of propaganda about the communes in the USSR. There were always films about the fantastic combine-harvesters with people singing on the back on their way to work. In the films there were always mountains and mountains of food. So many films showed how happy life was on the collective farms." Groups of Chinese peasants were sent on tours of Ukraine and Kazakhstan to visit "model" collectives and see how they worked. They noted that there was always lots of food on the table and modern equipment to work the fields. Mao Zedong decreed that China would adopt the same approach.

He started by establishing a state monopoly on grain. Grain was to be sold to the state at a fixed low price, ensuring that it could be sold abroad at a profit to raise money to pay for industrialization. Markets were closed, production quotas were assigned in each region, and a system of rationing was introduced to distribute grain in the cities. The state gradually took control of the grain supply. Mao then embarked on a collectivization program in order to increase production. Small groups of households, then dozens at a time, and finally hundreds at a time were combined to form collective farming communities. Tools, animals, and grain had to be pooled. This system was imposed by inviting farmers in a particular area to a meeting, and then not allowing them to leave until they "agreed" to form a collective—a process that sometimes took several days. As in the Soviet Union, a system of internal passports was introduced in 1956 to stop farmers fleeing to the cities.

Mao was following the Stalinist model closely, with predictably similar consequences. Grain production fell by 40 percent in 1956 alone, as collectivization robbed farmers of any incentive to maximize their output. People in some areas began to starve. Animals were killed and eaten, so that there were fewer of them to work the land. Meanwhile the Communist Party boasted of its great success in collectivizing agriculture. The harvest figures for 1949 were revised downward, to make subsequent years' figures look bigger, but food production had in fact fallen to a level below that of the 1930s. But Mao wanted to outdo the Soviet Union, and he began planning a "Great Leap Forward" that would, he hoped, industrialize China almost overnight. When some of his colleagues argued for a more gradual approach, he purged them from the Party. Even Nikita Krushchev, the new Soviet leader, who had come to power after Stalin's death in 1953, warned Mao not to go ahead with his program, which Krushchev understood was intended to "impress the world—especially the socialist world—with his genius and leadership." Krushchev was aware of the harm that Stalin's agricultural policies had done, and had quietly unwound many of them. But the growing rivalry between the Soviet Union and China meant that Mao did not just want to emulate Stalin's supposed achievements, but to outdo them. He promised that food production would double or triple within a year, along with the output of steel.

To make this happen, Party officials ordered the establishment of backyard furnaces and told everybody to hand over a certain quota of metal items. These would be transformed into steel in the furnaces, and the resulting metal would be used to mechanize agriculture. But steelmaking is rather more complicated than Mao realized. Large numbers of trees were cut down to fuel the furnaces, which merely turned perfectly good pots and pans into worthless pig iron. This unpleasant truth was kept from Mao by those in his inner circle. He was shown a backyard furnace that was seemingly producing high-quality steel, but the steel had actually been made elsewhere.

Mao's understanding of agriculture was even more tenuous than his grasp of metallurgy. In order to boost agricultural yields, the other main component of his Great Leap Forward, Mao drew up his own list of instructions for farmers, based largely on the barmy theories of Trofim Lysenko, a Soviet pseudoscientist. Mao advocated dense planting of seeds (which meant the soil could not sustain them), deep plowing (which damaged the fertility of the soil), greater use of fertilizer (but without chemicals, so household rubbish and broken glass was used instead), concentrating production on a smaller area of land (which quickly exhausted the soil), pest control (killing rats and birds, which caused the population of insects to explode), and increased irrigation (though the small dams and reservoirs that were constructed, being made of earth, soon collapsed).

Party officials, fearing for their own positions, went along with all this and pretended that Mao's instructions had resulted in amazing improvements in yields. Across China, bizarre achievements were announced: the growth of giant vegetables, and the crossbreeding of sunflowers with artichokes, tomatoes with cotton, and even sugarcane with maize and sorghum. Photographs were faked of miracle crops and plots where wheat had grown so densely that children could sit on top of its stalks. (The plants were actually transplanted into the field, and the children were sitting on a concealed table.) On one occasion peasants were told to transplant rice plants to fields along the route that Mao was traveling, to give the impression of an abundant crop; on another occasion vegetables were piled up by the roadside so that he could be told that peasants had abandoned them, having grown so much food that they had more than they could eat.

Mao was told that the grain harvest for 1958, the first after the launch of the Great Leap Forward, had doubled; in some cases yields in particular fields were said to have increased over 150-fold. Officials who could see what was really happening dared not question these claims. Where possible, farmers had ignored Mao's crackpot list of instructions, and the harvest was not much worse than that of previous years. But the redeployment of farmers in the misguided effort to make steel meant that not all the crops were gathered, and a lot of food rotted in the fields. Official figures said the harvest had doubled, however, so the procurements of grain demanded by the state's central granaries were much larger than in previous years. As different provinces vied to outdo each other in apparent productivity, they submitted larger and larger deliveries. Exports of grain doubled, providing apparent proof of China's agricultural miracle to the outside world. And in the autumn of 1958 Chinese farmers were told that there was abundant food, and that they could eat as much as they wanted in the communal kitchens. They did so, and by winter there was no food left.

People began to starve in large numbers. One Party leader later estimated that twenty-five million people were starving in early 1959. Mao refused to believe that the vast appropriations of grain being made by the state were causing shortages. If some regions were unable to meet their quotas, he said, it was because farmers were hiding their food. "We must recognize that there is a severe problem because production teams are hiding and dividing grain and this is a common problem all over the country," he declared. When some officials tried to explain the situation, Mao responded that if there were a few problems in some areas, those were "tuition fees that must be paid to gain experience." Peng Dehuai, the defense minister, who came from a peasant background and had experienced famine in his youth, accused Mao of sacrificing human lives in the pursuit of impossible production targets. He was stripped of his rank, placed under house arrest, and later exiled. Mao came to regard any reports of food shortages as personal attacks on his leadership, and he became even more determined to press ahead with his program. This meant that those officials who knew what was really going on became even less inclined to try to intervene.

Even higher grain-production targets were set for 1959. The harvest was about one-fifth smaller than in 1958, but officials reported another year of record yields, and to make their claims stand up they set about procuring all the grain they could find for delivery to the central government. (The state procurement quota was set at 40 percent in many areas, and 40 percent of the fictitious and vastly inflated harvest figures meant that in practice the entire harvest was seized.) When their quotas could not be met, even by seizing everything, officials began to search for hidden supplies of food that did not exist, as had previously happened in the Soviet Union. Perhaps the worst atrocities occurred in the province of Henan, where Party officials beat, tortured, and murdered thousands of peasants who were supposedly hiding grain. Some were set on fire; others had their ears cut off, were frozen to death, or were worked to death on construction projects. But there really was no food. People tried to eat grass and tree bark, and there were many cases of cannibalism.

By the end of 1959 millions of rural Chinese were starving. The communal kitchens served watery soup made of grass and anything else that could be found. As the crisis deepened, China cut itself off from the outside world. Relations with the Soviet Union were broken off so that Krushchev would not learn of the disaster. When problems were admitted, they were blamed on natural causes such as drought, but even then officials continued to insist that food was abundant and the people were happy. Mao began planning another big increase in production targets for 1960. But in much of the country the people were too weak to plant anything. Those in the cities suffered less; they were given grain rations from the central granaries, and thus were the last to be affected by the spread of the famine. In the countryside, Party officials had the first claim on what little food was available, so that many of them failed to realize the extent of the catastrophe on the land. Most of those who starved to death were peasants in rural communes.

Famine and starvation were widespread by the end of 1960, but Mao refused to recognize the problem. Senior members of the Communist leadership realized they had to act, if only to preserve the regime. They began to gather evidence to present to Mao and convince him of the scale of the disaster. But in some cases they were thwarted by local officials, loyal to Mao, who went to great lengths to deceive them; in other cases senior officials dared not confront Mao with the evidence, because they feared being punished for disloyalty. Hu Yaobang, one senior official, spent a sleepless night before an audience with Mao, wondering what to say. "I did not dare tell the Chairman the truth," he later admitted. "If I had done so this would have spelled the end of me. I would have ended up like Peng Dehuai."

In some areas senior Party officials managed to install local leaders who were prepared to reverse Mao's collectivization and get agriculture going again, by granting small plots to peasant households for their own use, as had previously been done in the Soviet Union. Collective kitchens were also dismantled, officials who had been dismissed for their opposition to collectivization were given their jobs back, and in some cases punishment was meted out to those who had brutally enforced Mao's policies. Deng Xiaoping, one of the reformers who had recognized that things had to change, famously declared at a meeting in March 1961 (at which Mao was not present) that "it does not matter whether a cat is black or white as long as it catches mice." In short, ideological considerations were less important than providing food.

But how could the reformers get Mao to agree to a retreat from collectivization, while enabling him to save face? Eventually, in mid-1961, Mao quietly agreed to allow the "lending" of some land to peasants so that they could grow their own food. But officially he refused to acknowledge that anything was wrong, or that anything had changed. Collective farming on communal fields continued, but in many parts of China people were also allowed to raise livestock and grow food on their own small plots on waste ground, and to trade in everything except grain (a fixed proportion of which still had to be handed over to the state). In Hunan this new policy came to be known as "save yourself production." Grain was shipped in from Australia and Canada, though it was sometimes repackaged in Chinese sacks to conceal its origin, since officially China was still reporting huge increases in grain yields.

The Great Leap Forward was a disaster that resulted in the worst famine in history. In all, some thirty to forty million died, though the full extent of the disaster only became apparent to the outside world in the 1980s, when American demographers analyzed population statistics released by China in 1979. Mao's agricultural policies, modeled on those of Stalin, caused overall grain yields to fall by 25 percent, and wheat yields by 41 percent. But the main cause of the famine was not inadequate food production so much as the farmers' lack of entitlement to it. The food they produced went to feed people in the cities, Party officials, and foreigners. During the crisis years China exported more than twelve million tons of grain and record amounts of pork, poultry, and fruit. Granaries in many parts of the country were well stocked, even as people starved. The famine was not caused by drought or flood, disease or pestilence. It was an entirely man-made disaster, the root cause of which was Mao's desire to use food to display the ideological superiority of Chinese socialism. Instead, he demonstrated precisely the opposite.

Food and the Collapse of the Soviet Union

What caused the collapse of the Soviet Union in 1991? According to Yegor Gaidar, a senior Russian politician who served in Boris Yeltsin's government in the era after the fall of the Soviet Union, the regime disintegrated in large part because it could not feed its people. The food crisis crept up on the Soviet Union over the course of several decades, but it had its roots in Stalin's industrialization program, back in the late 1920s. The leadership's obsession with industrial transformation meant that farm workers were less highly valued than industrial workers, and received much lower wages. As a result, those in the countryside took any opportunity they could to move to the cities and take a job in industry. As the urban population expanded, agricultural productivity stagnated.

When Nikita Khrushchev came to power after Stalin's death in 1953, he observed that grain yields had fallen by one fifth since 1940. As more of the shrinking food supply went to feed the growing urban population, there was less grain left over for export, so threatening

the industrialization program. The Soviet Union found itself between the closing jaws of a trap: The food demands of its urban population were growing, and supply could not keep up. What could be
done? One solution was to pay farmers more for their produce and
give them incentives to increase output. But that would have been
tantamount to reversing the collectivization program—a huge political U-turn. So instead Khrushchev decided to boost agriculture by
bringing virgin land under cultivation, and by paying the farmers
who worked on it the higher wages granted to industrial workers. Existing farmers' wages were left unchanged.

For a while, everything seemed to be going well. Grain production
increased for the first few years. But then it leveled off. Even with the
new land, the amount of food being produced per head of population
was still lower than it had been in 1913, and the state grain reserves
actually declined between 1953 and 1960. The new initiative had not
solved the problem. So the Soviet leadership tried another tack: boosting agricultural output by investing in tractors, combine harvesters,
and other equipment. Agricultural output did grow slowly in the 1960s
and 1970s, but consumption grew faster still. A turning point came in
1963, when the Soviet Union stopped exporting food and grain to its
satellite states in Eastern Europe—payments that had helped to maintain stability and political support in these satellites. Instead it bought
foreign grain, using 372 tons of gold—more than a third of the country's gold reserves—to pay for it. This was humiliating. Khrushchev
told his comrades it was vital to build up grain reserves again. "We
must have a year's supply of grain in seven years," he said. "The Soviet
regime cannot bear such shame again."

At the time, the need to resort to grain imports was blamed on a
one-off poor harvest in 1963. But there was a deeper problem. Much
of the newly cultivated land turned out to be in regions where the
size of the harvest was heavily dependent on the weather. During
the early 1970s imports and exports were roughly in balance, but by
the early 1980s the Soviet Union had become dependent on food imports, and by the mid-1980s it had become the world's largest grain

importer by a considerable margin—despite having been the world's largest exporter at the beginning of the twentieth century. It had to agree to long-term contracts to buy grain, guaranteeing annual purchases of nine million tons a year from the United States, five million from Canada, and four million from Argentina. The Soviet Union resorted to foreign loans, hard-currency reserves, and gold reserves (in particularly bad years) to pay for these imports. But this was not sustainable. Nor was exporting manufactured goods an option; most Soviet-made goods could not compete with those made in other countries. The Soviet Union had tried to industrialize using the proceeds from huge grain exports, but in the process it had undermined its agricultural productivity, a vital source of wealth.

Food prices continued to rise, and shortages became more widespread. Employees of government agencies and the military were allowed to buy food at reduced prices in special shops that were not open to the public. By 1981, according to Gaidar, "the USSR's political leadership was trapped, with no way out. It was impossible to speed up agricultural production sufficiently to meet the growing demand." The exploitation of oil reserves helped for a while. But the Soviets overexploited their oil fields for short-term gain, reducing their long-term prospects. High oil prices from the mid-1970s helped to pay for food imports, and for military spending to keep up with the United States. But the Soviet leaders assumed that oil prices would remain high indefinitely, and therefore they did not build up their foreign-currency reserves before the oil price fell sharply in 1985–86. Indeed, the Soviet Union's borrowing increased.

The Soviet leaders were all too aware of the danger of relying on their Cold War adversaries for food. But they had little choice. Mikhail Gorbachev, who came to power as the leader of the Soviet Union in 1985, began to introduce economic reforms, but to little avail as infighting paralyzed the regime. Soon all of the Soviet Union's oil revenue was being consumed by interest payments; and poor global grain harvests in 1989–90 drove up prices, in particular of wheat. The Soviet Union began to miss payments to foreign suppliers for food imports, causing some shipments to be halted. Many foodstuffs and consumer items became hard to find in shops; lengthy lines for sugar, butter, rice, salt, and other basic foods became commonplace.

On March 31, 1991, one of Gorbachev's aides wrote in his diary: "Yesterday the Security Council met on the food issue . . . more concretely, bread . . . In Moscow and other cities there are lines like the ones two years ago for sausage. If we don't get it somewhere, there may be famine by June. Of the republics, only Kazakhstan and Ukraine can (barely) feed themselves. That there is bread in the country turns out to be a myth. We scraped the bottom of the barrel to find hard currency and credit to buy it abroad. But we are no longer credit worthy . . . I drove around Moscow . . . the bakeries are padlocked or terrifyingly empty. I don't think Moscow has seen anything like this in all its history—even in the hungriest years." By this time many of the individual republics of the Soviet Union, starting with the Baltic states, and followed by Moldova, Ukraine, Belorussia, and Russia, had declared themselves sovereign states. Food shortages were a major cause of social unrest and of a collapse of the Soviet government's authority. "It remains difficult to ensure the presence of bread and other foodstuffs in a number of regions," noted the deputy minister of the interior. "Long lines form outside stores, the citizens criticize the local and central authorities in strong language, and some of them call for protest actions."

In autumn of 1991, an official memo reported: "The low harvest and the inability to expand imports, together with the refusal of farms to turn over their grain to the state, may put the country and the republic on the brink of famine. The only way out of this situation is to allow the farms to sell grain freely at market prices with further liberalization of retail prices for bread. Without a transition to free pricing in conjunction with an accelerated reduction of state control in agriculture and trade, there will be no incentive for growth in production." Finally, the penny had dropped. The Soviet policies of centralizing control of agriculture and controlling prices had failed. The only way forward, politicians conceded, was free trade and liberalization—in other words, capitalism. By this time the Soviet Union's disintegration was well advanced, and it formally ceased to exist on December 26, 1991, dissolving into its constituent states.

The Democracy of Food

Is it a coincidence that the worst famine in history happened in a Communist state? Not according to Amartya Sen, an Indian economist who won the Nobel prize in Economics in 1998. In his view, the combination of representative democracy and a free press makes famines much less likely to occur. "In the terrible history of famines in the world, no substantial famine has ever occurred in any independent and democratic country with a relatively free press," he wrote in 1999.

> We cannot find exceptions to this rule, no matter where we look: the recent famines of Ethiopia, Somalia, or other dictatorial regimes; famines in the Soviet Union in the 1930s; China's 1958–61 famine with the failure of the Great Leap Forward; or earlier still, the famines in Ireland or India under alien rule. China, although it was in many ways doing much better economically than India, still managed (unlike India) to have a famine, indeed the largest recorded famine in world history: Nearly 30 million people died in the famine of 1958–61, while faulty governmental policies remained uncorrected for three full years. The policies went uncriticized because there were no opposition parties in parliament, no free press, and no multiparty elections. Indeed, it is precisely this lack of challenge that allowed the deeply defective policies to continue even though they were killing millions each year.

Famines, Sen pointed out, are often blamed on natural disasters. But when such disasters strike democracies, politicians are more likely to act, if only to maintain the support of voters. "Not surprisingly, while India continued to have famines under British rule right up to independence (the last famine, which I witnessed as a child, was in 1943, four years before independence), they disappeared suddenly with the establishment of a multiparty democracy and a free press," Sen wrote.

The rise of democracy, which Sen calls "the preeminent development" of the twentieth century, would therefore explain why the use of food as an ideological weapon, like its use as a military weapon, has become much less widespread. A rare but striking example, at the time of writing in mid-2008, is its use by Robert Mugabe, Zimbabwe's dictator. He has presided over a collapse of Zimbabwe's agriculture, which has turned the country from a regional breadbasket into a disaster area. Between 2000 and 2008 agricultural output fell by 80 percent, unemployment increased to 85 percent, inflation rose to more than 100,000 percent, life expectancy fell below forty, and three million Zimbabweans, or about one fifth of the population, fled the country. With Zimbabwe in crisis, Mugabe maintained his grip on power through violence and intimidation, by rigging a series of elections, and by channeling food aid to members of his government and regions where his support was strongest, while denying it to people in areas known to be sympathetic to the opposition.

In June 2008 Mugabe was accused of offering food to people in opposition areas only if they gave up the identification documents needed to vote in the presidential election, to prevent them voting for the opposition candidate. A spokesman for the U.S. State Department, Sean McCormack, told reporters that Mugabe was "using food as a weapon, using the hunger of parents' children against them to prevent them from voting their conscience for a better kind of Zimbabwe." Mugabe responded that it was Western aid agencies that were using food for political ends, and he banned them from distributing food in opposition areas. "These western-funded NGOs also use food as a political weapon with which to campaign against government, especially in the rural areas," he said.

The overt use of food as a weapon in this way is now mercifully rare. In Western democracies, however, food has found another, more subtle political role. It is no longer a weapon, but has instead become a battlefield on which broader political fights take place. This is a consequence of the variety of food now available to Western consumers as a result of global trade, growing interest in the consequences and politics of food choices, and food's unusual status as a consumer product that acts as a lightning rod for broader social concerns. For almost any political view you want to express, there is a relevant foodstuff to buy or avoid.

Concerns over the environment can therefore be expressed by advocacy of local and organic products; "fair-trade" products aim to highlight the inequity of global-trade rules and the buying power of large corporations, while also funding social programs for low-paid workers and their families; arguments about genetically modified foodstuffs give expression to worries over the unfettered march of new technologies, and the extent to which farmers have become dependent on large agribusinesses. Shoppers can buy dolphin-friendly tuna, bird-friendly coffee, and bananas that support educational programs for growers in Costa Rica. They can express a desire for reconciliation in the Middle East by buying "peace oil" made in olive groves where Israelis and Palestinians work side by side. They can signal opposition to large companies by boycotting supermarkets in favor of small shops or farmers' markets.

Food can also be used to make specific protests against companies or governments. In 1999 when José Bové, a French political activist, wanted to express his opposition to the might of the United States and to the impact of multinational corporations on French traditions and local companies, he did so by dismantling a McDonald's restaurant in the town of Millau, loading the rubble onto tractors, and dumping it outside the town hall. More recently, in South Korea in 2008 there were huge public protests against American beef imports, ostensibly on safety grounds; but the protests really gave voice to broader unease about the removal of trade barriers and to concerns that South Korea's ruling party was allowing itself to be pushed around by the country's superpower patron.

The idea of using food to make wider political points can be traced back to 1791, when British consumers who wanted to express their opposition to slavery began to boycott sugar. A stream of pamphlets ensued, including the Anti-Saccharine Society's deliberately shocking manifesto, illustrated with a cross-section of a slave ship to show how tightly the shackled men were packed into it. A newspaper advertisement placed by James Wright, a Quaker merchant, in 1792 was representative of the mood: "Therefore being impressed with the Sufferings and Wrongs of that deeply-injured People, and also with an Apprehension, that while I am a Dealer in that Article, which appears to be a principal support of the Slave Trade, I am encouraging Slavery, I take this Method of informing my Customers, that I mean to discontinue selling the Article of Sugar till I can procure it through Channels less contaminated, more unconnected with Slavery, and less polluted with Human Blood."

Campaigners claimed that if just thirty-eight thousand British families stopped buying sugar, the impact on the planters' profits would be severe enough to bring the trade to an end. At the boycott's peak, one of the leaders of the campaign claimed that three hundred thousand people had given up sugar. Some campaigners smashed teacups in public, since they were tainted by sugar. Tea parties became social and political minefields. It was a faux pas to ask for sugar if it was not offered by an abstaining hostess. But not all sugar was equally bad. Some people regarded more expensive sugar from the East Indies to be less ethically problematic—until it transpired that it, too, was very often grown by slaves. When the slave trade was abolished by Britain in 1807, it was unclear whether the boycott, or a series of slave revolts, had made the most difference. Some even argued that the boycott had made things worse: As planters' profits fell, they might well have treated their slaves even more cruelly. But there was no doubt that the sugar boycott had drawn attention to the slavery question and helped to mobilize political opposition.

The same is true of today's food debates. Their real significance lies not so much in their direct impact, but in the way in which they can provide a leading indicator to governments about policy, and encourage companies to change their behavior. Food has a unique political power, for several reasons: food links the world's richest consumers with its poorest farmers; food choices have always been a potent means of social signaling; modern shoppers must make dozens of food choices every week, providing far more opportunities for political expression than electoral politics; and food is a product you consume, so eating something implies a deeply personal endorsement of it. But there are limits to its power. Real change—such as abolishing slavery in the nineteenth century, or overhauling world trade or tackling climate change today—ultimately requires political action by governments. Voting with your food choices is no substitute for voting at the ballot box. But food provides a valuable arena in which to debate difference choices, a mechanism by which societies indicate what they feel strongly about, and a way to mobilize broader political support. Those in positions of power, whether in politics or business, would be foolish to ignore such signals.

PART VI

Food, Population,
and Development

II

FEEDING THE WORLD

[Agriculture's] principal object consists in the production of
nitrogen under any form capable of assimilation.

—JUSTUS VON LIEBIG, 1840

THE MACHINE THAT CHANGED THE WORLD

Compared with the flight of Wright brothers' first plane or the deto-
nation of the first atomic bomb, the appearance of a few drips of colorless liquid at one end of an elaborate apparatus in a laboratory in
Karlsruhe, Germany, on a July afternoon in 1909 does not sound very
dramatic. But it marked the technological breakthrough that was to
have arguably the greatest impact on mankind during the twentieth
century. The liquid was ammonia, and the tabletop equipment had
synthesized it from its constituent elements, hydrogen and nitrogen.
This showed for the first time that the production of ammonia
could be performed on a large scale, opening up a valuable and
much-needed new source of fertilizer and making possible a vast
expansion of the food supply—and, as a consequence, of the human
population.

The link between ammonia and human nutrition is nitrogen. A
vital building block of all plant and animal tissue, it is the nutrient
reponsible for vegetative growth and for the protein content of cereal grains, the staple crops on which humanity depends. Of course,
plants need many nutrients, but in practice their growth is limited
by the availability of the least abundant nutrient. Most of the time
this is nitrogen. For cereals, nitrogen deficiency results in stunted

growth, yellow leaves, reduced yields, and low protein content. An abundance of available nitrogen, by contrast, promotes growth and increases yield and protein content. Nitrogen compounds (such as proteins, amino acids, and DNA) also play crucial roles in the metabolisms of plants and animals; nitrogen is present in every living cell. Humans depend on the ingestion of ten amino acids, each built around a nitrogen atom, to synthesize the body proteins needed for tissue growth and maintenance. The vast majority of these essential amino acids comes from agricultural crops, or from products derived from animals fed on those crops. An inadequate supply of these essential amino acids leads to stunted mental and physical development. Nitrogen, in short, is a limiting factor in the availability of mankind's staple foods, and in human nutrition overall.

The ability to synthesize ammonia, combined with new "high-yield" seed varieties specifically bred to respond well to chemical fertilizers, removed this constraint and paved the way for an unprecedented expansion in the human population, from 1.6 billion to 6 billion, during the course of the twentieth century. The introduction of chemical fertilizers and high-yield seed varieties into the developing world, starting in the 1960s, is known today as the "green revolution." Without fertilizer to nourish crops and provide more food—increasing the food supply sevenfold, as the population grew by a factor of 3.7—hundreds of millions of people would have faced malnutrition or starvation, and history might have unfolded very differently.

The green revolution has had far-reaching consequences. As well as causing a population boom, it helped to lift hundreds of millions of people out of poverty and underpinned the historic resurgence of the Asian economies and the rapid industrialization of China and India—developments that are transforming geopolitics. But the green revolution's many other social and environmental side effects have made it hugely controversial. Its critics contend that it has caused massive environmental damage, destroyed traditional farming practices, increased inequality, and made farmers dependent on expensive seeds and chemicals provided by Western companies. Doubts have also been expressed about the long-term sustainability of chemically inten-sive farming. But for better or worse, there is no question that the green revolution did more than just transform the world's food supply in the second half of the twentieth century; it transformed the world.

The Mystery of Nitrogen

The origins of the green revolution lie in the nineteenth century, when scientists first came to appreciate the crucial role of nitrogen in plant nutrition. Nitrogen is the main ingredient of air, making up 78 percent of the atmosphere by volume; the rest is mostly oxygen (21 percent), plus small amounts of argon and carbon dioxide. Nitrogen was first identified in the 1770s by scientists investigating the proper-ties of air. They found that nitrogen gas was mostly unreactive and that animals placed in an all-nitrogen atmosphere suffocated. Yet having learned to identify nitrogen, the scientists also discovered that it was abundant in both plants and animals and evidently had an im-portant role in sustaining life. In 1836 Jean-Baptiste Boussingault, a French chemist who took a particular interest in the chemical foun-dations of agriculture, measured the nitrogen content of dozens of substances, including common food crops, various forms of manure, dried blood, bones, and fish waste. He showed in a series of experi-ments that the effectiveness of different forms of fertilizer was di-rectly related to their nitrogen content. This was odd, given that atmospheric nitrogen was so unreactive. There had to be some mech-anism that transformed nonreactive nitrogen in the atmosphere into a reactive form that could be exploited by plants.

Some scientists suggested that lightning created this reactive ni-trogen by breaking apart the stable nitrogen molecules in the air; others speculated that there might be trace quantities of ammonia, the simplest possible compound of nitrogen, in the atmosphere. Still others believed that plants were somehow absorbing nitrogen from the air directly. Boussingault took sterilized sand that contained no nitrogen at all, grew clover in it, and found that nitrogen was then present in the sand. This suggested that legumes such as clover could somehow capture (or "fix") nitrogen from the atmosphere directly. Further experiments followed, and eventually in 1885 another French chemist, Marcelin Berthelot, demonstrated that uncultivated soil was also capable of fixing nitrogen, but that the soil lost this ability if it was sterilized. This suggested that nitrogen fixation was a property of something in the soil. But if that was the case, why were leguminous plants also capable of fixing nitrogen?

The mystery was solved by two German scientists, Hermann Hellriegel and Hermann Wilfarth, the following year. If nitrogen-fixing was a property of the soil, they reasoned, it should be transferable. They put pea plants (another kind of legume) in sterilized soil, and they added fertile soil to some of the pots. The pea plants in the sterile soil withered, but those to which fertile soil had been added flourished. Cereal crops, however, did not respond to the application of soil in the same way, though they did respond strongly to nitrate compounds. The two Hermanns concluded that the nitrogen-fixing was being done by microbes in the soil and that the lumps, or nodules, that are found on the roots of legumes were sites where some of these microbes took up residence and then fixed nitrogen for use by the plant. In other words, the microbes and the legumes had a cooperative, or symbiotic, relationship. (Since then, scientists have discovered nitrogen-fixing microbes that are symbiotic with freshwater ferns and supply valuable nitrogen in Asian paddy fields; and nitrogen-fixing microbes that live in sugarcane, explaining how it can be harvested for many years from the same plot of land without the use of fertilizer.)

Nitrogen's crucial role as a plant nutrient had been explained. Plants need nitrogen, and certain microbes in the soil can capture it from the atmosphere and make it available to them. In addition, legumes can draw upon a second source of nitrogen, namely that fixed by microbes accommodated in their root nodules. All this explained how long-established agricultural practices, known to maintain or replenish soil fertility, really worked. Leaving land fallow for a year or two, for example, gives the microbes in the soil a chance to replenish the nitrogen. Farmers can also replenish soil nitrogen by recycling various forms of organic waste (including crop residues, animal manures, canal mud, and human excrement), all of which contain small amounts of reactive nitrogen, or by growing leguminous plants such as peas, beans, lentils, or clover.

These techniques had been independently discovered by farmers all over the world, thousands of years earlier. Peas and lentils were being grown alongside wheat and barley in the Near East almost from the dawn of agriculture. Beans and peas were rotated with wheat, millet, and rice in China. In India, lentils, peas, and chickpeas were rotated with wheat and rice; in the New World, beans were interleaved with maize. Sometimes the leguminous plants were simply plowed back into the soil. Farmers did not know why any of this worked, but they knew that it did. In the third century B.C., Theophrastus, the Greek philosopher and botanist, noted that "the bean best reinvigorates the ground" and that "the people of Macedonia and Thessaly turn over the ground when it is in flower." Similarly, Cato the Elder, a Roman writer of the second century B.C., was aware of beneficial effects of leguminous crops on soil fertility, and he advised that they should "be planted not so much for the immediate return as with a view to the year later." Columella, a Roman writer of the first century A.D., advocated the use of peas, chickpeas, lentils, and other legumes in this way. And the "Chhi Min Yao Shu," a Chinese work, recommended the cultivation and plowing-in of adzuki beans, in a passage that seems to date from the first century B.C. Farmers did not realize it at the time, but growing legumes is a far more efficient way to enrich the soil than the application of manure, which contains relatively little nitrogen (typically 1 to 2 percent by weight).

The unraveling of the role of nitrogen in plant nutrition coincided with the realization, in the mid-nineteenth century, of the imminent need to improve crop yields. Between 1850 and 1900 the population in western Europe and North America grew from around three hundred million to five hundred million, and to keep pace with this growth, food production was increased by placing more land under cultivation on America's Great Plains, in Canada, on the Russian steppes, and in Argentina. This raised the output of wheat and maize, but there was a limit to how far the process could go. By the early twentieth century there was little remaining scope for placing more land under cultivation, so to increase the food supply it would be necessary to get more food per unit area—in other words, to increase yields. Given the link between plant growth and the availability of nitrogen, one obvious way to do this was to increase the supply of nitrogen. Producing more manure from animals would not work, because animals need food, which in turn requires land. Sowing leguminous plants to enrich the soil, meanwhile, means that the land cannot be used to grow anything else in the meantime. So, starting as early as the 1840s, there was growing interest in new, external sources of nitrogen fertilizer.

Solidified bird excrement from tropical islands, known as guano, had been used as fertilizer on the west coast of South America for centuries. Analysis showed that it had a nitrogen content thirty times higher than that of manure. During the 1850s, imports of guano went from zero to two hundred thousand tons a year in Britain, and shipments to the United States averaged seventy-six thousand tons a year. The Guano Islands Act, passed in 1856, allowed American citizens to take possession of any uninhabited islands or rocks containing guano deposits, provided they were not within the jurisdiction of any other government. As guano mania took hold, entrepreneurs scoured the seas looking for new sources of this valuable new material. But by the early 1870s it was clear that the guano supply was being rapidly depleted. ("This material, though once a name to conjure with, has now not much more than an academic interest, owing to the rapid exhaustion of supplies," observed the *Encyclopaedia Britannica* in 1911.) Instead, the focus shifted to another source of nitrogen: the huge deposits of sodium nitrate that had been discovered in Chile. Exports boomed, and in 1879 the War of the Pacific broke out between Chile, Peru, and Bolivia over the ownership of a contested nitrate-rich region in the Atacama Desert. (Chile prevailed in 1883, depriving Bolivia of its coastal province, so that it has been a land-locked country ever since.)

Even when the fighting was over, however, concerns remained over the long-term security of supply. One forecast, made in 1903, predicted that nitrate supplies would run out by 1938. It was wrong—there were in fact more than three hundred years of supply, given the consumption rate at the time—but many people believed it. And by this time sodium nitrate was in demand not only as a fertilizer, but also to make explosives, in which reactive nitrogen is a vital ingredient. Countries realized that their ability to wage war, as well as their ability to feed their populations, was becoming dependent on a reliable supply of reactive nitrogen. Most worried of all was Germany. It was the largest importer of Chilean nitrate at the beginning of the twentieth century, and its geography made it vulnerable to a naval blockade. So it was in Germany that the most intensive efforts were made to find new sources of reactive nitrogen.

One approach was to derive it from coal, which contains a small amount of nitrogen left over from the biomass from which it originally formed. Heating coal in the absence of oxygen causes the nitrogen to be released in the form of ammonia. But the amount involved is tiny, and efforts to increase it made little difference. Another approach was to simulate lightning and use high voltages to generate sparks that would turn nitrogen in the air into more reactive nitrous oxide. This worked, but it was highly energy-intensive and was therefore dependent on the availability of cheap electricity (such as excess power from hydroelectric dams). So imported Chilean nitrate remained Germany's main source of nitrogen. Britain was in a similarly difficult situation. Like Germany, it was also a big importer of nitrates, and was doing its best to extract ammonia from coal. Despite efforts to increase agricultural production, both countries relied on imported wheat.

In a speech at the annual conference of the British Association for the Advancement of Science in 1898, William Crookes, an English chemist and the president of the association, highlighted the obvious solution to the problem. A century after Thomas Malthus had made the same point, he warned that "civilised nations stand in deadly peril of not having enough to eat." With no more land available, and with concern growing over Britain's dependence on wheat imports, there was no alternative but to find a way to increase yields. "Wheat preeminently demands nitrogen," Crookes observed. But there was no scope to increase the use of manure or leguminous plants; the supply of fertilizer from coal was inadequate; and by relying on Chilean nitrate, he observed, "we are drawing on the Earth's capital, and our drafts will not perpetually be honoured." But there was an abundance of nitrogen in the air, he pointed out—if only a way could be found to get at it. "The fixation of nitrogen is vital to the progress of civilised humanity," he declared. "It is the chemist who must come to the rescue . . . it is through the laboratory that starvation may ultimately be turned into plenty."

A Productive Dispute

In 1904 Fritz Haber, a thirty-six-year-old experimental chemist at the Technische Hochschule in Karlsruhe, was asked to carry out some research on behalf of a chemical company in Vienna. His task was to determine whether ammonia could be directly synthesized from its constituent elements, hydrogen and nitrogen. The results of previous experiments had been unclear, and many people thought direct synthesis was impossible. Haber himself was skeptical, and he replied that the standard way to make ammonia, from coal, was known to work and was the easiest approach. But he decided to go ahead with the research anyway. His initial experiments showed that nitrogen and hydrogen could indeed be coaxed into forming ammonia at high temperature (around 1,000 degrees Centigrade, or 1,832 degrees Fahrenheit) in the presence of an iron catalyst. But the proportion of the gases that combined was very small: between 0.005 percent and 0.0125 percent. So although Haber had resolved the

Fritz Haber.

question of whether direct synthesis was possible, he also seemed to have shown that the answer had no practical use.

And there things might have rested, had it not been for Walther Hermann Nernst, another German chemist, who was professor of physical chemistry at Göttingen. Although he was only four years older than Haber, Nernst was a more eminent figure who had made contributions in a number of fields. He had invented a new kind of light bulb, based on a ceramic filament, and an electric piano with guitar-style pickups, though neither was a commercial success. Nernst was best known for having proposed a "heat theorem" (now known as the third law of thermodynamics) in 1906 that would win him the Nobel prize in Chemistry in 1920. This theorem could be used to predict all sorts of results, including the proportion of ammonia that should have been produced by Haber's experiment. The problem was that Nernst's prediction was 0.0045 percent, which was below the range of possible values determined by Haber. This was the only anomalous result of any significance that disagreed with Nernst's theory, so Nernst wrote to Haber to point out the discrepancy. Haber performed his original experiment again, obtaining a more precise answer: This time around the proportion of ammonia produced was 0.0048 percent. Most people would have regarded that as acceptably close to Nernst's predicted figure, but for some reason Nernst did not. When Haber presented his new results at a conference in Hamburg in 1907, Nernst publicly disputed them, suggested that Haber's experimental method was flawed, and called upon Haber to withdraw both his old and new results.

Haber was greatly distressed by this public rebuke from a more senior scientist, and he suffered from digestion and skin problems as a result. He decided that the only way to restore his reputation was to perform a new set of experiments to resolve the matter. But during the course of these experiments he and his assistant, Robert Le Rossignol, discovered that the ammonia yield could be dramatically increased by performing the reaction at a higher pressure, but a lower temperature, than they had used in their original experiment. Indeed, they calculated that increasing the pressure to 200 times atmospheric pressure, and dropping the temperature to 600 degrees Centigrade (1,112 degrees Fahrenheit), ought to produce an ammonia yield of 8 percent—which would be commercially useful. The dispute with Nernst seeemed trivial by comparison and was swiftly forgotten, and Haber and Le Rossignol began building a new apparatus that would, they hoped, produce useful amounts of ammonia. At its center was a pressurized tube just 75 centimeters tall and 13 centimeters in diameter, surrounded by pumps, pressure gauges, and condensers. Haber refined his apparatus and then invited representatives of BASF, a chemical company that was by this time funding his work, to come and see it in operation.

The crucial demonstration took place on July 2, 1909, in the presence of two employees from BASF, Alwin Mittasch and Julius Kranz. During the morning a mishap with one of the bolts of the high-pressure equipment delayed the proceedings for a few hours. But in the late afternoon the apparatus began operating at 200 atmospheres and about 500 degrees Centigrade, and it produced an ammonia yield of 10 percent. Mittasch pressed Haber's hand in excitement as the colorless drops of liquid ammonia began to flow. By the end of the day the machine had produced 100 cubic centimeters of ammonia. A jubilant Haber wrote to BASF the next day: "Yesterday we began operating the large ammonia apparatus with gas circulation in the presence of Dr. Mittasch and were able to keep its production uninterrupted for about five hours. During this whole time it functioned correctly and it continuously produced liquid ammonia. Because of the lateness of the hour, and as we all were tired, we stopped the production because nothing new could be learned from continuing the experiment."

Ammonia synthesis on a large scale suddenly seemed feasible. BASF gave the task of converting Haber's benchtop apparatus into a large-scale, high-pressure industrial process to one of its senior

Fritz Haber's experimental apparatus.

chemists, Carl Bosch. He had to work out how to generate the two feedstock gases (hydrogen and nitrogen) in large quantities and at low cost; to find suitable catalysts; and, most difficult of all, to develop large steel vessels capable of withstanding the enormous pressures required by the reaction. The first two converters built by Bosch, which were around four times the size of Haber's apparatus, failed when their high-pressure reaction tubes exploded after around eighty hours of operation, despite being encased in reinforced concrete. Bosch realized that the high-pressure hydrogen was weakening the steel tubes by depleting them of the carbon that gives steel its strength and resilience. After much trial and error he redesigned the inside of the tubes to prevent this problem. His team also developed new kinds of safety valves to cope with the high pressures and temperatures; devised clever heat-exchange systems to reduce the energy required by the synthesis process; and built a series of small converters to allow large numbers of different materials to be tested as possible catalysts. Bosch's converters gradually got bigger during 1910 and 1911, though they were still producing only a few kilograms of ammonia per day. Only in February 1912 did output first exceed one ton in a single day.

By this time Haber and BASF were under attack from rivals who were contesting Haber's patents on the ammonia-synthesis process. Chief among them was Walther Nernst, whose argument with Haber had prompted Haber to develop the new process in the first place. Some of Haber's work had built on earlier experiments by Nernst, so BASF offered Nernst an "honorarium" of ten thousand marks a year for five years in recognition of this. In return, Nernst dropped his opposition to Haber's patents, and all other claims against Haber were subsequently thrown out by the courts.

Meanwhile ever-larger converters, now capable of producing three to five metric tons a day, were entering service at BASF's new site at Oppau. These combined Haber's original methods with Bosch's engineering innovations to produce ammonia—from nitrogen in the air, and hydrogen extracted from coal—using what is now known as

Carl Bosch.

the Haber-Bosch process. By 1914 the Oppau plant was capable of producing nearly 20 metric tons of ammonia a day, or 7,200 metric tons a year, which could then be processed into 36,000 metric tons of ammonium sulphate fertilizer. But the outbreak of the First World War in August 1914 meant that much of the ammonia produced by the plant was soon being used to make explosives, rather than fertilizer. (Germany's supply of nitrate from Chile was cut off after a series of naval battles, in which the British prevailed.)

The war highlighted the way in which chemicals could be used both to sustain life or to destroy it. Germany faced a choice between using its new source of synthetic ammonia to feed its people or supply its army with ammunition. Some historians have suggested that without the Haber-Bosch process, Germany would have run out of nitrates by 1916, and the war would have ended much sooner. German production of ammonia was scaled up dramatically after 1914, but with much of the supply being used to make munitions, maintaining food production proved to be impossible. There were widespread food shortages, contributing to the collapse in morale that preceded

Germany's defeat in 1918. So the synthesis of ammonia prolonged the war, but Germany's inability to produce enough for both munitions and fertilizer also helped to bring about the war's end.

Haber himself strikingly embodies the conflict between the constructive and destructive uses of chemistry. During the war he turned his attention to the development of chemical weapons, while Bosch concentrated on scaling up the output of ammonia. Haber oversaw the first successful large-scale use of chemical weapons in April 1915, when Germany used chlorine gas against the French and Canadians at Ypres, causing some five thousand deaths. Haber argued that killing people with chemicals was no worse than killing them with any other weapon; he also believed that their use "would shorten the war." But his wife, Clara Immerwahr, who was a chemist herself, violently disagreed, and she shot herself using her husband's gun in May 1915. Scientists of many nationalities protested when Haber was awarded the 1918 Nobel prize in Chemistry, in recognition of his pioneering work on the synthesis of ammonia and its potential application in agriculture. The Royal Swedish Academy of Sciences, which awarded the prize, commended Haber for having developed "an exceedingly important means of improving the standards of agriculture and the well-being of mankind." This was a remarkably accurate prediction, given the impact that fertilizers made using Haber's process were to have in subsequent decades. But the fact remains that the man who made possible a dramatic expansion of the food supply, and of the world population, is also remembered today as one of the fathers of chemical warfare.

When scientists in Britain and other countries had tried to replicate the Haber-Bosch process themselves during the war, they had been unable to do so because crucial technical details had been omitted from the relevant patents. These patents were confiscated after the war, and BASF's plants were scrutinized by foreign engineers, leading to the construction of similar plants in Britain, France, and the United States. During the 1920s the process was refined so that it could use methane from natural gas, rather than coal, as the source of hydrogen. By the early 1930s the Haber-Bosch process had overtaken Chilean nitrates to become the dominant source of artificial fertilizer, and global consumption of fertilizer tripled between 1910 and 1938. Having relied on soil microbes, legumes, and manure for thousands of years, mankind had decisively taken control of the nitrogen cycle. The outbreak of the Second World War prompted the construction of even more ammonia plants to meet the demand for explosives, which meant that there was even more fertilizer-production capacity available after the war ended in 1945. The stage was set for a further dramatic increase in the use of artificial fertilizer. But if its potential to increase food production was to be exploited to the full, new seed varieties would also be needed.

The Rise of the Dwarfs

The availability of artificial fertilizer allowed farmers to supply much more nitrogen to their crops. For cereals such as wheat, maize, and rice, this produced larger, heavier seed heads, which in turn meant higher yields. But now that they were no longer constrained by the availability of nitrogen, farmers ran into a new problem. As the use of fertilizer increased the size and weight of the seed heads, plants became more likely to topple over (something farmers call "lodging"). Farmers had to strike a balance between applying plenty of fertilizer to boost yield, but not so much that the plants' long stalks were unable to support the seed heads. The obvious solution was to switch to short, or "dwarf" varieties with shorter stalks. As well as being able to support heavier seed heads without lodging, dwarf varieties do not waste energy growing a long stalk, so more energy can be diverted to the seed head. They therefore boost yield in two ways: by allowing more fertilizer to be applied, and by turning applied nutrients more efficiently into useful grain, rather than useless stalk.

During the nineteenth century, dwarf varieties of wheat, probably descended from a Korean variety, had been developed in Japan. They greatly impressed Horace Capron, the United States' commissioner of agriculture, who visited Japan in 1873. "No matter how much manure is used . . . on the richest soils and with the heaviest of yields, the wheat stalks never fall down and lodge," he noted. In the early twentieth century these Japanese dwarf varieties were crossed with varieties from other countries. One of the resulting strains, Norin 10, was a cross between Japanese wheat and two American varieties. It was developed in Japan, at the Norin breeding station, and was transferred to the United States after the Second World War. Norin 10 had unusually short, strong stems (roughly two feet tall, rather than three feet), and responded well to heavy applications of nitrogen fertilizer. But it was susceptible to disease, so agronomists in different countries began to cross it with local varieties in order to combine Norin 10's dwarf characteristics with the pest resistance of other varieties. This led to new, high-yielding varieties of wheat suitable for use in particular parts of the world. In industrialized countries where use of nitrogen fertilizer was growing quickly, the new varieties descended from Norin 10 made possible an impressive increase in yield. By this time new, high-yielding varieties of maize had also become widespread, so that during the 1950s the U.S. secretary of agriculture complained that the country was accumulating "burdensome surpluses" of grain that were expensive to store.

When it came to the developing world, one man did more than anyone else to promote the spread of the new dwarf varieties: Norman Borlaug, an American agronomist. He went to Mexico in 1944 at the behest of the Rockefeller Foundation, which had established an agricultural research station there to help to improve poor crop yields. The foundation had concluded that boosting yields was the most effective way to provide agricultural and economic assistance, and reduce Mexico's dependence on grain imports. Borlaug was put in charge of wheat improvement, and his first task was to develop varieties that were resistant to a disease called stem rust, which was a particular problem in Mexico at the time: It reduced Mexico's wheat harvest by half between 1939 and 1942. Borlaug created hundreds of crossbreeds of local varieties, looking for strains that demonstrated good resistance to stem rust and also provided strong yields. Within a few years he had produced new, resistant breeds with yields 20 to 40 percent higher than the traditional varieties in use in Mexico.

Mexico was an excellent place to carry out such research, Borlaug realized, because one wheat crop could be grown in the highlands in the summer, and another in the lowland desert in the winter. He developed a new system called "shuttle breeding," in which he carried the most promising results from one end of the country to another. This broke the traditional rule that plants should only be bred in the area in which they would subsequently be planted, but it sped up the

Norman Borlaug.

breeding process, since Borlaug could produce two generations a year rather than one. His rule-bending also had another, unanticipated benefit: In order to thrive as both summer and winter crops, the resulting varieties could not afford to be fussy about the difference in the number of hours of daylight between the two seasons. This meant their offspring could subsequently be cultivated in a wide range of different climates.

In 1952 Borlaug heard about the work being done with Norin 10, and the following year he received some seeds from America. He began to cross his new Mexican varieties with Norin 10, and with a new variety that had been created by crossing Norin 10 with an American wheat called Brevor. Within a few years he had developed new wheat strains with insensitivity to day length and good disease resistance that could, with the use of nitrogen fertilizer, produce more than twice the yield of traditional Mexican varieties. Borlaug wanted to make further improvements, but curious farmers visiting his research station were taking samples of his new varieties and planting them, and they were spreading fast. So Borlaug released his new seeds in 1962. The following year, 95 percent of Mexico's wheat was based on one of Borlaug's new varieties, and the wheat harvest was six times larger than it had been nineteen years earlier when he had first arrived in the country. Instead of importing 200,000 to 300,000 tons of wheat a year, as it had done in the 1940s, Mexico exported 63,000 tons of wheat in 1963.

Following the success of his new high-yielding dwarf wheat varieties in Mexico, Borlaug suggested that they could also be used to improve yields in other developing countries. In particular, he suggested India and Pakistan, which were suffering from poor harvests and food shortages at the time and had become dependent on foreign food aid. Borlaug's suggestion was controversial, because it would mean encouraging farmers to grow wheat rather than indigenous crops. Borlaug maintained, however, that since wheat produced higher yields and more calories, his new dwarf wheat varieties presented a better way for South Asian farmers to take advantage of the advent of cheap nitrogen fertilizer than trying to increase yields of indigenous crops. Monkombu Sambasivan Swaminathan, an Indian geneticist who was an adviser to the agriculture minister, invited Borlaug to visit India, and Borlaug arrived in March 1963 and began promoting the use of his Mexican wheat. Some small plots were planted, and they produced impressive results at the following year's wheat harvest: With irrigation and the application of nitrogen fertilizer, the yields were around five times that of local Indian varieties, which typically produced around one ton per hectare. Swaminathan later recalled that "when small farmers, who with the help of scientists organised the National Demonstration Programme, harvested over five tons of wheat per hectare, its impact on the minds of other farmers was electric. The clamour for seeds began."

Another impressive harvest in early 1965 prompted the Indian government to order 250 tons of seed from Mexico for further trials. But wider adoption of the new seeds was held back by political and bureaucratic objections. A turning point came when the monsoon, which normally occurs between June and September, failed in 1965. This caused grain yields to fall by nearly one fifth and made India even more dependent on foreign food aid. The government sent officials to Mexico to place an order for eighteen thousand tons of the new wheat seeds—enough to sow around 3 percent of India's wheat-growing areas. As the ship carrying the seeds departed for Bombay, war broke out between India and Pakistan, diverting attention from the food crisis gripping the region. And by the time the seeds were being unloaded in September, it was apparent that the monsoon had failed for a second year.

The combination of political instability, population growth, and drought in South Asia gave rise to a new outbreak of Malthusianism in the late 1960s. Across the developing world, the population was growing twice as fast as the food supply. Pundits predicted imminent disaster. In their 1967 book *Famine—1975!*, William and Paul Paddock argued that some countries, including India, Egypt, and Haiti, would simply never be able to feed themselves and should be left to starve. That same year, one fifth of the United States' wheat harvest was shipped to India as emergency food aid. "The battle to feed all of humanity is over," declared Paul Ehrlich in his 1968 best-seller, *The Population Bomb*. He predicted that "in the 1970s and 1980s hundreds of millions of people will starve to death in spite of any crash programs embarked upon now." He was particularly gloomy about India, declaring that it "couldn't possibly feed two hundred million more people by 1980."

As with Thomas Malthus's predictions nearly two centuries ear-lier, the technologies that would disprove these gloomy predictions were already quietly spreading. Following the introduction of Bor-laug's high-yield varieties from Mexico, wheat yields in India in-creased from twelve million tons in 1965 to nearly seventeen million tons in 1968 and twenty million in 1970. The harvest in 1968 was so large that schools had to be closed in some areas so that they could be used for grain storage. India's grain imports fell almost to zero by 1972, and the country even became an exporter for a while during the 1980s. Further improvements in yields followed in subsequent years as Indian agronomists crossed the Mexican varieties with local strains to improve disease resistance. India's wheat harvest reached 73.5 million tons in 1999.

Norman Borlaug's early success with high-yield dwarf varieties of wheat, meanwhile, had inspired researchers to do the same with rice. The International Rice Research Institute (IRRI), based in the Philip-pines and funded by the Rockefeller and Ford foundations, was es-tablished in 1960. Borlaug's shuttle-breeding approach was adopted to speed up the development of new varieties. As with wheat, re-searchers took dwarf varieties, many of them developed in Japan, and crossed them with the local varieties planted in other countries. In 1966 researchers at the IRRI created a new variety, called IR8, by crossing a Chinese dwarf variety (itself derived from a Japanese strain) with an Indonesian strain called Peta. At the time, traditional strains of rice produced yields of around one ton per hectare. The new variety produced five tons without fertilizer, and ten tons when fertilizer was applied. It became known as "miracle rice" and was quickly adopted throughout Asia. IR8 was followed by further dwarf strains that were more disease resistant and matured faster, making it possible to grow two crops a year for the first time in many regions.

In a prescient speech in March 1968, William Gaud of the United States Agency for International Development had highlighted the impact that high-yield varieties of wheat were starting to have in Pakistan, India, and Turkey. "Record yields, harvests of unprecedented size and crops now in the ground demonstrate that throughout much of the developing world—and particularly in Asia—we are on the verge of an agricultural revolution," he said. "It is not a violent red revolution like that of the Soviets, nor is it a white revolution like that of the Shah of Iran. I call it the green revolution. This new revolution can be as significant and as beneficial to mankind as the Industrial Revolution of a century and a half ago." The term "green revolution" immediately gained widespread currency, and it has remained in use ever since.

The impact of the green revolution was already apparent by 1970, and in that year Norman Borlaug was awarded the Nobel Peace Prize. "More than any other single person of this age, he has helped to provide bread for a hungry world," the Nobel Committee declared. He had "turned pessimism into optimism in the dramatic race between population explosion and our production of food." In his acceptance speech, Borlaug pointed out that the increase in yields was due not simply to the development of dwarf varieties, but to the combination of the new varieties with nitrogen fertilizer. "If the high-yielding dwarf wheat and rice varieties are the catalysts that have ignited the green revolution, then chemical fertilizer is the fuel that has powered its forward thrust," he said.

In the three decades after 1970, the new high-yield dwarf varieties of wheat and rice swiftly displaced traditional varieties across the developing world. By 2000, the new seed varieties accounted for 86 percent of the cultivated area of wheat in Asia, 90 percent in Latin America, and 66 percent in the Middle East and Africa. Similarly, the new varieties of rice accounted for 74 percent of the rice-producing area across Asia in 2000, and 100 percent in China, the world's largest rice producer. As well as offering increased yields—provided appropriate fertilizers and irrigation were available—they also increased cereal production in other, indirect ways. Farmers switched to wheat and rice from other crops, and farmers who were already growing wheat and rice could, in some cases, grow more than one crop a year by switching to new varieties. All this increased cereal production and meant that the food supply grew faster than the population. Asia's population increased by 60 percent between 1970 and 1995, but cereal production in the region over the same period more than doubled. Overall, nitrogen fertilizer has supported around four billion people born in the century since Haber's demonstration in 1909. By 2008, nitrogen fertilizer was responsible for feeding 48 percent of the world's population. Haber-Bosch nitrogen sustains more than three billion people, nearly half of humanity. They are the offspring of the green revolution.

12

PARADOXES OF PLENTY

Accelerated agricultural progress is the best safety net against
hunger and poverty, because in most developing countries over
70 percent of the population depend on agriculture for their
livelihood.

—M. S. SWAMINATHAN, 2004

THE RESURGENCE OF ASIA

To appreciate the impact of the green revolution, it helps to take a
long-term view of world economic activity. The big picture is that for
most of human history, most people were poor. Before 1700, average
per-capita income was low, roughly constant over time, and varied
very little between countries. Of course, some people in each country
were fabulously rich. But the average income was remarkably consistent: One calculation puts it at the equivalent of five hundred dollars
per annum (measured in 1990 dollars) for most of the world for the
past two millennia. Today, however, there are wide variations between
countries. Britain was the first to experience a "growth takeoff" when
it began the process of industrialization in the eighteenth century. It
was soon followed by other western European nations and Europe's
offshoots (the United States, Canada, Australia, and New Zealand).
By 1900, their average per-capita income was ten times higher than
that in Asia or Africa. Some countries are now rich, and others poor,
because industrialization took place in the rich countries first; the poor
countries are those in which it took place much later, or has not happened at all. So why does industrialization start at different times and

proceed at different rates? It is one of the most fundamental questions in developmental economics.

The answer has a lot to do with agricultural productivity. Poor countries cannot embark on economic development until they can meet their subsistence needs. They find themselves trapped in what economists have called a state of "high food drain" in which most of the population is tied down in inefficient agricultural production. Normally, when a particular activity is inefficient, people switch to other things. But agriculture is a special case: Food is vital, so people have no choice but to stick with farming, even when productivity is low. Indeed, low productivity means that more resources must be devoted to agriculture in order to maintain production. This is sometimes called the "food problem." To escape from this trap, a country must experience an improvement in agricultural productivity, so that the food supply expands more quickly than the population. This then allows some of the population to switch to higher-value industrial activities without worrying about where their food is going to come from. The proportion of the population engaged in agriculture shrinks as agricultural productivity improves, and industrialization is under way. This is what happened in Britain in the eighteenth century, as a series of improvements in agriculture liberated workers from the land and allowed industry to take root. Industrial goods could then be traded for food imports, further accelerating the switch from agriculture to industry. For all this to happen, the right infrastructure and market conditions must be in place. But a surge in agricultural productivity is essential to kick-start the process; no country has been able to industrialize without one. (The two exceptions are Singapore and Hong Kong, city-states that did not have significant agricultural sectors in the first place.)

Another notable feature of world economic history is that for most of human history, Asia was the wealthiest region on earth. In 1 A.D. it is estimated that Asia accounted for 75 percent of world economic output. This is not because people in Asia were individually richer; average per-capita income was, after all, remarkably consistent from one part of the world to another. It is because there were more people in Asia than in other regions, in large part because rice agriculture supports higher population densities. But Asia's share of world economic output began to decline with the rise of the western European economies in the second millennium A.D. By 1700, western Europe accounted for more than 20 percent of world output, and Asia's share had fallen below 60 percent. The crossover came in the late nineteenth century, as European nations industrialized and grew wealthier, and kept much of Asia under their colonial thumbs. By around 1870, Europe's share of world output had risen to 35 percent, and Asia's share had declined to approximately the same level, despite its much larger population. The rapid industrialization of the United States meant that by 1950, the United States and western Europe each accounted for around 25 percent of world output, and Asia's share (excluding Japan) had fallen to 15 percent.

But in the final quarter of the twentieth century something remarkable happened, and the tables were turned. Rapid growth in several Asian countries pushed the region's share of world output back up to 30 percent, ahead of western Europe or the United States. Economic output per capita more than doubled between 1978 and 2000 in India, and it increased nearly fivefold in China. Asia is now home to the world's fastest-growing economies and has reclaimed its historical position as the wealthiest region, measured by share of world output. Its rapid growth in the past few years—sometimes referred to as Asia's economic miracle—has created wealth more rapidly than at any time in history, and has lifted hundreds of millions of people out of poverty. Many observers now expect China's economy to surpass that of the United States in size by 2035, making China the world's leading economic power. Just as the twentieth century was dominated by the rise of the United States, the twenty-first looks set to be the Asian century, dominated by the rise of China. But this is arguably just a return to the ancient status quo, after a brief interlude in which European powers and their colonial offshoots briefly stole the limelight.

The resurgence of Asia has many causes, but it would not have been possible without the dramatic increase in agricultural productivity triggered by the green revolution. Between 1970 and 1995, cereal production in Asia doubled, the number of available calories per person increased by 30 percent, and the prices of wheat and rice fell. The immediate impact of agricultural progress is to reduce poverty, for the simple reason that the poor are most likely to work in agriculture, and food accounts for the majority of their household spending. Sure enough, the proportion of Asia's population living in poverty fell from around 50 percent in 1975 to 25 percent in 1995. The absolute number of Asians in poverty also declined, from 1.15 billion in 1975 to 825 million in 1995, even though the population increased by 60 percent. Agricultural progress also put Asia on the path toward economic development and industrialization.

For growth in agricultural productivity to translate into broader economic growth and industrialization, however, several other things need to happen. Farmers must have incentives to increase production; there must be infrastructure in place to transport seeds and chemicals onto farms, and food off them; and there must be adequate access to credit to enable farmers to purchase seeds, fertilizer, tractors, and so forth. Agricultural progress can trigger sudden economic growth, but how quickly it happens depends crucially on nonagricultural reforms being introduced at the same time. Consider the examples of China and India.

After the failure of the Great Leap Forward, reformers within the Chinese government took a more conventional approach to increasing agricultural output, and arranged to buy five medium-size ammonia plants from Britain and the Netherlands between 1963 and 1965. Once they were up and running, these plants supplied 25 percent of the nitrogen applied to China's fields. But the upheaval of the Cultural Revolution in the mid-1960s meant that by 1972, per-capita food output was still lower than it had been in the 1950s, and rapid population growth meant the amount of agricultural land available per person was shrinking fast. The only option was to increase yields.

U.S. president Richard Nixon visited China in 1972, opening up trade between the two countries, and the first deal signed was an order for thirteen of America's largest and most modern fertilizer plants—the biggest purchase of its kind in history. Within a few years China had overtaken the United States to become the world's leading consumer of fertilizer, and then became the biggest producer. China also quickly adopted the new high-yield dwarf varieties of wheat and rice.

But policy reforms were needed, too. After Mao's death in 1976, reformers led by Deng Xiaoping concluded that agriculture was the bottleneck preventing further economic progress. They introduced a "two-tier" system in which households were allocated land and could decide what to cultivate on it. Provided they met a state quota of around 15 to 20 percent of their output, they could sell the rest and keep the proceeds. This provided farmers with incentives to increase production, and it proved very successful in the areas where it was first tested, so that it was introduced throughout China between 1979 and 1984. The targets and quotas were gradually removed, and this approach was then adopted as a model for the rest of the Chinese economy, in which free enterprise was allowed alongside the state sector, and quickly outgrew it. As agriculture became more productive, rural workers were able to move into other areas, starting with food processing and distribution, and gradually expanding into other industries and services. By the mid-1990s, rural "town and village enterprises," almost none of which existed in 1978, accounted for 25 percent of the Chinese economy. These firms began to put pressure on state-run companies in the cities, which were less competitive. This in turn prompted broader economic reforms, the establishment of special economic zones for industrial activity, efforts to attract foreign investment, and so on—all of which fueled further economic growth. The result was an astonishing reduction in poverty, from 33 percent of the population in 1978 to 3 percent in 2001.

India was slower to introduce the policy reforms needed to allow improvements in agricultural productivity to translate into broader economic growth. Instead, India's main concern was agricultural self-sufficiency, and to this end the agricultural sector was tightly regulated and controlled by the government, with price controls, restrictions on the movement of agricultural goods within the country, and barriers that served to discourage foreign trade. With the adoption of green-revolution technology and investment in infrastructure for irrigation, agricultural output expanded, farmers' incomes rose, and nonagricultural employment increased. Falling food prices benefited the poor more broadly, so that the percentage of the rural population in poverty fell from 64 percent in 1967 to 34 percent by 1986. In 1986 the wheat harvest was forty-seven million tons, half of which was set aside as a reserve. The following year, when the monsoon failed, leading to the worst drought of the century, India was able to feed itself without loss of life, and without relying on outside aid.

This was a clear demonstration that India had achieved its goal of self-sufficiency in food. Liberalization of the manufacturing sector began in 1991, and India entered a period of rapid growth. The proportion of the population in poverty declined from 55 percent in 1973 to 26 percent in 2000. Some forecasters expect India to become the world's third-largest economy, after China and the United States, by 2035. But India has been less successful than China in promoting the creation of rural jobs outside agriculture, the crucial step that enables the poor to participate in wider economic growth. Food production, distribution, and retailing are still highly regulated. The proportion of the population involved in agriculture remains high, and there is widespread concern about inequality. The green revolution set the stage for India's rapid development, but further reforms are needed if the benefits are to be more widely distributed.

The Ghost of Malthus

A second long-term consequence of the green revolution has been its impact on global demographics—the size and structure of the population. Once again, it makes sense to take a historical step back. In

3000 B.C., as the first civilizations were emerging, the world population was a mere ten million or so, or roughly the population of London today. By 500 B.C., as Greece entered its Golden Age, the world population had increased to one hundred million. It was not until 1825, some ten thousand years after the dawn of agriculture, that the human population first reached one billion. It took another century to reach two billion, in 1925; and a mere thirty-five years to reach three billion, in 1960. The rapid growth of the world population was likened at the time to an explosion, and led to dire predictions of imminent famine. But the expansion in the food supply made possible by the green revolution meant that the population continued to climb, reaching four billion in 1975, five billion in 1986, and six billion in 1999. The fifth billion was added in a mere eleven years; the sixth billion in a further thirteen. The population is expected to reach seven billion in 2012, after a further thirteen years, according to the United States Census Bureau. In retrospect, then, it is clear that the population-growth rate has now started to slow.

Does population growth drive food production, or vice versa? Demographers have argued it both ways. A burgeoning population creates incentives to find new ways to increase the food supply; but greater availability of food also means that women are more fertile, and children are healthier and more likely to survive. So there is no simple answer. But history clearly shows that in cases where the greater availability of food enables a country to industrialize, there is a population boom, followed by a fall in the population-growth rate as people become wealthier—a phenomenon called "demographic transition."

In a preindustrial society, it makes sense to have as many children as possible. Many of them will not survive, due to disease or malnutrition. But once those that do survive are old enough to work in the fields, they can produce more food than they consume, so the household will benefit overall (provided that availability of labor is the main constraint on agricultural production). Having lots of children also provides security in old age, when parents expect to be looked after by their offspring. In such preindustrial societies, both birth rates and death rates are very high, and the population grows slowly. This was the situation for most of human history.

The advent of new farming techniques, crops, and tools that boost food output then move the society into a second phase in which the population grows quickly. This is what happened in western Europe starting in the eighteenth century, following the introduction of maize and potatoes from the New World and the spread of new farming practices. In this phase, the birth rate remains high but the death rate falls, resulting in a population boom. At the same time, greater agricultural productivity means that a smaller proportion of the population is needed in farming, opening the way to urbanization and industrialization.

This in turn seems to cause people to reassess their attitude to having children: Wealth, it seems, is a powerful contraceptive. The decline in infant mortality means parents in rural areas do not need to have so many children in order to be sure of having enough people to work in the fields, or to look after them in old age. In urban areas, meanwhile, parents may take the view that it makes sense to have a smaller number of children, given the cost of housing, clothing, and educating them. This is sometimes characterized as a switch from emphasizing child "quantity" to child "quality." In addition, as female literacy improves and women enter the workforce, they may delay marriage and change their attitude toward childbearing. And governments in industrializing countries generally introduce reforms banning child labor and making education compulsory, which means that children are a drain on household resources until they reach working age. The result is that the birth rate falls, and the population stabilizes. This pattern can be clearly seen in Western nations, which were the first to industrialize. In some European countries the fertility rate (the average number of births per woman) has now fallen below the replacement rate. Most developing countries, meanwhile, are now in the midst of their demographic transition.

Of course, the reality is more complicated than this simple model suggests, due to other factors such as the effects of migration, the impact of HIV/AIDS in Africa, and China's one-child policy, introduced in 1980. But having initially sustained a population boom, the green revolution is now tipping many countries, and consequently the world as a whole, into demographic transition. According to forecasts published by the United Nations in 2007, the world population is expected to reach eight billion around 2025, and to peak at 9.2 billion in 2075, after which it will decline.

Research carried out in the village of Manupur, in the Indian Punjab, illustrates how the demographic transition has manifested itself on the ground. In 1970, men in the village all said that they wanted as many sons as possible. But when researchers returned to the village in 1982, following the introduction of green-revolution crops, fewer than 20 percent of men said that they wanted three or more sons, and contraceptives were being widely used. "These rapid changes in family size preference and contraceptive practice are indications that the demographic transition will continue, if not accelerate, in rural areas experiencing the green revolution," the researchers concluded. Similarly, Bangladeshi women had an average of seven children in 1981. Following the widespread adoption of green-revolution technologies in the 1980s and the rapid expansion of the country's textiles industry in the 1990s, however, that figure has fallen to an average of two or three.

The world will face new challenges as its population shrinks—not least the difficulty of looking after an infirm and aging population, which is already a concern in developed countries where the fertility rate has fallen. But the peak of world population may now be in sight. Once the population starts to decline, worries about population growth outstripping food supply may start to seem rather old-fashioned. A flood of bestselling books will no doubt warn of the dangers of the coming population implosion. But the ghost of Malthus will finally have been laid to rest.

PROBLEMS WITH THE GREEN REVOLUTION

New technologies often have unforeseen consequences, and the technologies of the green revolution are no exception. High-yield seed varieties, which require artificial fertilizers, other agricultural chemicals, and large amounts of water, have caused environmental problems in many parts of the world. Nitrogen-laden runoff from agricultural land has created "dead zones" in some coastal areas, stimulating the growth of algae and weeds and reducing the amount of oxygen in the water and thereby affecting fish and shellfish populations. In some cases high-yield varieties proved to be less resistant than traditional varieties to pests or diseases. This necessitated a greater use of pesticides, overuse of which can contaminate the soil and harm beneficial insects and other wildlife, reducing biodiversity. Pesticides can also cause health problems for farm workers. According to the World Health Organization, pesticides cause around one million cases of acute unintentional poisoning a year and are also involved in around two million suicide attempts, leading to some 220,000 deaths a year. (The availability of pesticides has made pesticide poisoning the most widespread method of suicide in the developing world.) A further worry is the depletion of water supplies. In the Punjab, the cradle of India's green revolution, for example, the proliferation of millions of tube wells caused the water table to fall by more than fifteen feet between 1993 and 2003 alone, and many farmers now have insufficient water to irrigate their crops.

Much can be done to mitigate these problems, however. More frugal and precise application of fertilizer can reduce runoff without affecting yields. Fertilizer intensity has in fact been declining in recent years in some developed countries. In the United States, maize yields have increased from 42 kilograms per kilogram of fertilizer in 1980 to 57 kilograms in 2000. Similar improvements have been achieved with wheat in Britain and rice in Japan. But in many developing countries fertilizer is heavily subsidized by governments, discouraging more efficient use. More can also be done to reduce the unnecessary use of pesticides and minimize harmful side effects. During the rollout of the green revolution, farmers were instructed that the use of pesticides was a necessary component of "modern" agriculture, which resulted in overuse. Some farmers were told to apply pesticides according to a calendar, whether or not such applications were necessary. The use of pesticides is now flat or declining, and biological pest-control techniques are being promoted in conjunction with chemicals, making the best use of both traditional and modern practices. This hybrid approach, called "integrated pest management," can reduce the use of pesticides by 50 percent for vegetable crops. In some cases it can eliminate the need for pesticides altogether in rice production, according to the United Nations' Food and Agricultural Organization.

Similarly, there is plenty of scope for improvements in water use. Far more attention is now being paid to aquifer management, for example, and to the deployment of rainwater harvesting and storage systems, and of more water-efficient irrigation systems such as drip-irrigation technology (which also reduces nitrogen runoff). Clearly defined water rights that can be traded by farmers can also encourage more sensible use of water, by encouraging farmers to concentrate on the most appropriate crops. It seems odd to grow water-intensive crops such as potatoes in Israel, oranges in Egypt, cotton in Australia, and rice in California, for example, when all of these crops could be grown more cheaply and efficiently elsewhere. And in the Punjab, the provision of free electricity to farmers, along with subsidies for growing rice, a water-intensive crop, encouraged farmers to leave their water pumps running continuously. In recent years, growing concern about the scarcity of water for agriculture—it has even been called the "oil of the twenty-first century"—has prompted policymakers to pay greater attention to the development of sensible water policies.

The environmental problems associated with high-yield farming must also be weighed against its unseen environmental benefits, in the form of damage to ecosystems that would otherwise have been done in order to increase food production. High-yield varieties have enabled food production to multiply with only a marginal increase in land use. Asia's cereal production doubled between 1970 and 1995, for example, but the total area cultivated with cereals increased by just 4 percent. Globally, the figures are even more striking. Norman Borlaug has pointed out that world output of cereal grains tripled between 1950 and 2000, but the area used for cereal cultivation increased by only 10 percent. Without green-revolution technologies, he contends, there would either have been mass starvation, or enormous amounts of virgin land (such as forests) would have had to have been taken under cultivation.

Many critics of the green revolution advocate a return to traditional, or organic, agricultural techniques that do not rely on chemical fertilizers and pesticides. This would reduce both the direct environmental impact of agriculture (in the form of nitrogen runoff and pesticide use) and its indirect impact (since the production of chemical fertilizer is an energy-intensive process that consumes natural gas and contribtues to climate change). But farming without the use of chemical fertilizers produces lower yields, so more land is then needed to provide the same amount of food. Studies have found that organic production of wheat, maize, and potatoes, for example, requires two or three times as much land as conventional production. Global agriculture in 1900, using almost no chemical fertilizer, supported about 1.6 billion people on an area of about 850 million hectares (2.1 billion acres), according to the University of Manitoba's Vaclav Smil, an expert on the nitrogen cycle. Farming using fertilizer-free (that is, organic) methods on today's 1,500 million hectares (3.7 billion acres) would support only 3.2 billion people on mostly vegetarian diets, he estimates, or half of today's global population.

That said, the use of fertilizer in the developed world could be reduced while still providing enough food to provide adequate nutrition, despite the fall in yields. That is because rich countries produce more food than they need, in part because paying subsidies to farmers encourages overproduction. The excess allows for unnecessarily protein-rich diets (resulting in rising levels of obesity in rich countries) and large exportable surpluses. So there is scope to switch some food production to less chemically intensive methods, such as organic farming. The situation in the developing world is very different, however. In rich countries, chemical fertilizer supplies only about 45 percent of the nitrogen applied to fields. But in poorer countries it supplies as much as 80 percent. It is the use of fertilizer that makes the difference between inadequate and adequate nutrition, and in many developing countries the supply of dietary protein remains inadequate even so.

By the late 1990s, 75 percent of all nitrogen being applied to crops in China was coming from chemical fertilizers. Since 90 percent of the protein consumed by Chinese is homegrown, this means that two thirds of the nitrogen in China's food comes from the Haber-Bosch process. Traditional methods, such as planting nitrogen-fixing legumes or using animal manure, simply cannot supply as much nitrogen per hectare. In many other populous developing countries, the level of food production now exceeds the level that could be produced by traditional, fertilizer-free methods. There may be scope to reduce the amount of fertilizer used by more precise application, but it is difficult to see how it can be eliminated altogether without reducing food output.

There are no easy answers. Both conventional and organic farming have environmental costs and benefits. During the twentieth century mankind became dependent on artificial nitrogen, and turning back the clock is not an option. Chemically intensive agriculture has undesirable environmental side effects, and more effort is undoubtedly needed to mitigate them. But the consequences to humanity of abandoning the green revolution would surely be far worse.

A Second Green Revolution?

Between January 2007 and April 2008, after several years of stable prices, wheat prices abruptly doubled, rice prices tripled, and maize prices increased 50 percent. For the first time since the early 1970s, food riots erupted in several countries simultaneously. In Haiti the prime minister was forced to resign by crowds of protesters chanting "We're hungry!" Two dozen people died in food riots in Cameroon. The president of Egypt mobilized the army and told soldiers to start baking bread. In the Philippines, a new law was introduced making the hoarding of rice punishable by life imprisonment. After years in which farmers and development specialists had lamented the low prices of staple foods, the era of cheap food seemed to have abruptly come to an end. In many respects the origins of this food crisis can be traced back to the consequences of the green revolution.

One consequence was that governments and aid agencies lost interest in agriculture as a means of promoting development. According to the World Bank, the proportion of "official development assistance" spent on agriculture fell from 18 percent in 1979 to 3.5 percent in 2004. There were several reasons for this shift, according to the World Bank's 2008 World Development Report. To some extent it seemed that the food problem had been solved. There were food gluts in North America and Europe, and low international prices for staple foods, the result of both green-revolution technologies and subsidies to farmers in the developed world. As a result, donors lost their enthusiasm for funding agricultural projects in the developing world. Waning investment by governments in agricultural research, starting in the 1990s, meant that growth in yields slowed.

Farmers and environmental groups in developed countries also convinced donors to reduce funding for agricultural development in the developing world. The farmers regarded developing countries as valuable export markets, and did not want their governments to fund potential competitors. And environmental groups highlighted the pollution caused by chemically intensive agriculture, and managed to discredit the green revolution in the eyes of many donors. In the 1980s, when Norman Borlaug began a campaign to extend the green revolution to Africa, where it had had little impact, he found that attitudes were changing. Environmental lobby groups had persuaded the World Bank and the Ford Foundation that promoting the use of chemical fertilizers in Africa was a bad idea.

The emergence of the Chinese and Indian middle classes, who could afford to eat more meat-rich, Western-style diets, increased demand for cereal grains for use as animal feed, raising prices. And the diversion of food crops into biofuel production also increased prices, though exactly how much impact this had on world prices is uncertain. Higher oil prices also contributed to higher food prices, by increasing production and transport costs and by raising the price of fertilizer (since the price of natural gas, from which fertilizer is made, is pegged to the price of oil). In short, although the supply of food continued to grow, the rate of growth declined (to 1 to 2 percent a year since the mid-1990s) and was unable to keep pace with the growth in demand (at around 2 percent a year). Tellingly, India started importing wheat again in 2006. Like many countries, India also banned the export of many foodstuffs in an effort to maintain supplies for the domestic population. Such export bans further increased international food prices, by reducing the amount of food available on global markets.

If nothing else, the food crisis has put agriculture back on the international development agenda, after years of neglect. In the short term, the appropriate response to the crisis is a rapid increase in humanitarian food aid. Policies promoting biofuels made from food crops must also be reconsidered. But in the medium term, shipping large quantities of food from rich to poor countries makes things worse, because it undermines the market for local producers. The long-term answer is to embark upon a new effort to increase agricultural production in the developing world, by placing renewed emphasis on agricultural research and the development of new seed varieties, investment in the rural infrastructure needed to support farmers, greater access to credit, the introduction of new crop-insurance schemes, and so on. All of this may sound rather familiar, because it is, in essence, a call for a second "green revolution."

Inevitably, this has revived the arguments about the pros and cons of the original green revolution. Some advocates of a second green revolution emphasize the potential of genetically modified seeds, now under development, that produce their own pesticides or are designed to make more efficient use of water and fertilizer. (This has been referred to as a "doubly green revolution.") Advocates of organic farming, meanwhile, regard the food crisis as an ideal opportunity to promote greater use of organic methods, particularly in Africa where yields are low. In much of Africa, raising yields even to the level of pre-fertilizer agriculture in other countries would be a valuable achievement.

Clearly, any new green revolution should take into account the lessons learned since the 1960s. There are many new techniques to draw upon that can improve yields while minimizing environmental problems. Some are low-tech, such as burying precisely measured pellets of fertilizer to minimize runoff, or using particular beetles and spiders to keep pests at bay. Seeds can be coated with fungicides or pesticides directly, reducing the need to spray chemicals. And a particularly promising approach is "conservation agriculture" (also known as "no till" or "conservation tillage" farming), a set of techniques developed since the 1970s that minimize the tilling of the soil, or even eliminate it altogether.

Farmers practicing conservation agriculture leave crop residues on their fields after harvest, rather than plowing them in or burning them off. Cover crops are then planted to protect the soil. (Planting legumes as cover crops helps to increase soil nitrogen.) In the spring, the cover crop and any weeds are either killed using a herbicide, or chopped up on the surface using special machinery. Planting of the main crop is then done using machines that guide seeds into slots in the soil below the protective layer of residue. All this helps to reduce soil erosion, since covered, unplowed soil is less likely to be washed or blown away. Water is used more efficiently because the soil's ability to hold water increases, and less water is lost to runoff or evaporation. Conservation agriculture also saves fuel and reduces energy consumption, since about half as many passes over the field using machinery are required. Less fertilizer is usually needed because less nitrogen is lost to the environment; this also reduces nitrogen pollution of waterways. Conservation agriculture is most widely used in North and South America, where it was first developed, but it still accounts for only a small proportion (around 6 percent) of cultivated land worldwide, so there is much potential to expand its use.

It is possible that new genetically modified seeds will deliver on their promise of more efficient nitrogen uptake and water use. New seeds are also being engineered to grow in soils that are too salty for traditional varieties. The development of such seeds will take several more years, and it is too early to say how successful they will be. It is certainly overstating the case to suggest that genetic modification is a "silver bullet" that will fix the world's various food problems. But it would be foolish to rule out its use altogether. At the same time, there may be organic techniques that can be more widely applied, particularly when it comes to biological pest control and growing crops in arid areas. Some studies show that organic methods may produce higher yields for some crops in dry conditions, for example.

To ensure an adequate supply of food as the world population heads toward its peak and climate change shifts long-established patterns of agriculture, it will be necessary to assemble the largest possible toolbox of agricultural techniques. Different methods will be the most appropriate in different regions. It may make sense to grow staple crops using chemically intensive methods in some parts of the world, and to trade them for specialist crops grown using traditional methods elsewhere, for example. It is far too simplistic to suggest that the world faces a choice between organic fundamentalism on the one hand and blind faith in biotechnology on the other. The future of food production, and of mankind, surely lies in the wide and fertile middle ground in between.

EPILOGUE

INGREDIENTS OF THE FUTURE

There is no feast which does not come to an end.

—Chinese proverb

On a remote island in the Arctic circle, seven hundred miles from the North Pole, an incongruous concrete wedge protrudes from the snow on the side of a mountain. Reflective steel, mirrors, and prisms, built into an aperture on its outside face, reflect the polar light during the summer months, making the building gleam like a gem set into the landscape. In the dark of the winter it glows with an eerie white, green, and turquoise light from two hundred optical fibers, ensuring that the building remains visible for miles around. Behind its heavy steel entrance doors, a reinforced-concrete tunnel extends 125 meters (410 feet) into the bedrock. And behind another set of doors and two airlocks are three vaults, each 27 meters long, 6 meters tall, and 10 meters wide (89 by 20 by 33 feet). These vaults will not store gold, works of art, secret blueprints, or high-tech weaponry. Instead they will store something far more valuable—something that is arguably mankind's greatest treasure. The vaults will be filled with billions of seeds.

The Svalbard Global Seed Vault, on the Norwegian island of Spitsbergen, is the world's largest and safest seed-storage facility. The seeds it contains are stored inside gray four-ply envelopes made of polyethylene and aluminum, packed into sealed boxes, and stacked on shelves in the three vaults. Each envelope holds an average of five hundred seeds, and the total capacity of the vault is 4.5 million envelopes, or more than two billion seeds. This is far larger than any existing seed bank: When the first vault is only half full, the Svalbard Global Seed Vault will be the world's largest collection of seeds.

The vault's careful design and positioning should also make it the world's safest collection. There are about 1,400 seed banks world-wide, but many of them are vulnerable to wars, natural disasters, or a lack of secure funding. In 2001, Taliban fighters wiped out a seed bank in Afghanistan that contained ancient types of walnut, almond, peach, and other fruits. In 2003, during the American invasion of Iraq, a seed bank in Abu Ghraib was destroyed by looters, and rare varieties of wheat, lentils, and chickpeas were lost. Much of the collection at the national seed bank in the Philippines was lost in 2006 when it was swamped by muddy water during a typhoon. A Latin American seed bank almost lost its collection of potatoes when its refrigerators broke down. Malawi's seed bank is a freezer in the corner of a wooden shack. Physical dangers aside, the funding for many seed banks is also precarious. Kenya's entire seed bank was almost lost because its administrators could not afford to pay the electricity bill. The Svalbard facility, which will act as a backup for all of these national seed banks, has been designed to minimize both man-made and natural risks, and its running costs will be paid by the Norwegian government, which also paid for its construction.

As well as being built in one of the most remote places on earth, the Svalbard vault is tightly secured with steel doors and coded locks, is monitored from Sweden by video-link, and is protected by motion detectors set up around the site. (Polar bears provide a further deterrent to intruders: People in the region are advised to carry a high-powered rifle whenever they venture outside a settlement.) The structure is built into a mountain that is geologically stable and has a low level of background radiation. And it is 130 meters (426 feet) above sea level, so it will remain untouched even under the most pessimistic scenarios for rising sea levels in the future. The vault's refrigeration system, powered by locally mined coal, will keep the seeds at −18 degrees Celsius (−0.3 degrees Fahrenheit). Even if the refrigeration system fails, the vault's position, deep below the permafrost, ensures that the inside temperature will never exceed −3.5 degrees Celsius (25.7 degrees Fahrenheit), which is cold enough to protect most of the seeds for many years. In normal operation, a few seeds from each sample will be withdrawn from time to time and planted, so that fresh seeds can be harvested. (Some seeds, such as lettuce seeds, can only be stored for about fifty years.) In this way, the thousands of varieties of seeds can be perpetuated almost indefinitely.

The purpose of the Svalbard vault is to provide an insurance policy against both a short-term threat and a long-term one. The short-term threat—the disruption of global agriculture by climate change—seems likely to be the next way in which food will influence the course of human progress. In many countries, climate change could mean that the coolest years in the late twenty-first century will be warmer than the hottest years of the twentieth century. The conditions in which today's common crop varieties were developed will no longer apply. William Cline, an expert on the economic impact of global warming at the Center for Global Development in Washington, D.C., predicts that climate change will reduce agricultural output by 10 to 25 percent by 2080 in developing countries unless action is taken. In some cases the impact is far more dramatic: India's food output could fall by 30 to 40 percent. Agricultural output in some developed countries, by contrast, which typically have lower average temperatures, may increase slightly as temperatures rise. The worst-case scenario is that there could be wars over food, as global shifts in agricultural production lead to widespread droughts and food shortages and provoke conflict over access to agricultural land and water supplies.

The more optimistic scenario is that agriculture can adapt to changes in the climate, which are inevitable to some degree even if mankind manages to reduce emissions of greenhouse gases dramatically during the course of the twenty-first century. As formerly rich agricultural land becomes too arid for farming and previously cold, damp areas become more suitable for agriculture, seeds with new characteristics will be needed. And that is where the Svalbard seed bank comes in. The spread of high-yield seed varieties, in the wake of the green revolution, means that many traditional crop varieties are no longer being planted, and are being lost. Of the 7,100 types of apple that were being grown in America in the nineteenth century, for example, 6,800 are now extinct. Globally, the United Nations' Food and Agriculture Organization estimates that around 75 percent of crop varieties were lost during the twentieth century, and further varieties are being lost at the rate of one a day. These traditional varieties very often produce lower yields than modern varieties, but collectively they represent a valuable genetic resource that must be preserved for use in the future.

Consider the case of a variety of wheat known as PI 178383. It was dismissed as "a hopelessly useless wheat" by Jack Harlan, an American botanist, when he collected a sample of it in Turkey in 1948. It did badly in cold winters, had a long, weak stalk that made it fall over easily, and was susceptible to a disease called leaf rust. But in 1963, when plant breeders were looking for a way to make American wheat resistant to another disease, called stripe rust, the supposedly useless Turkish wheat turned out to be invaluable. Tests showed that it was immune to four kinds of stripe rust and forty-seven other wheat diseases. It was crossbred with local varieties, and today nearly all the wheat grown in the Pacific Northwest is descended from it. Harlan's seed collecting trips, in which he traveled simply, often on a donkey, had gathered priceless genetic material. There is, in short, no way to tell which varieties will turn out to be useful in the future for their drought tolerance, immunity to disease, or pest resistance. So the logical thing to do is to conserve as many seeds as possible as securely as possible—which is what the Svalbard facility is designed to do.

It also provides insurance against a longer-term threat. Someday a nuclear war, an asteroid striking the earth, or some other global calamity might make it necessary to rebuild human civilization from scratch, starting with its deepest foundation: agriculture. Some of the seeds being stored at Svalbard are capable of surviving for millennia, even if its refrigeration systems fail. Wheat seeds can last 1,700 years, barley seeds for 2,000 years, and sorghum seeds for 20,000 years. Perhaps, hundreds of years from now, an intrepid band of explorers will head to Svalbard to retrieve the crucial ingredients needed to restart the process that first began in the Neolithic period, some 10,000 years ago.

Despite the Svalbard seed bank's futuristic design and high-tech features, there is an echo of the Neolithic in its purpose: to store seeds safely. It was the ability to store seeds as an insurance policy against future food shortages that first led people to take a particular interest in cereal crops. This started them down the path to domestication, farming, and all the other consequences that have been described in this book. From the dawn of agriculture to the green revolution, food has been an essential ingredient in human history. And whether the seeds stored at Svalbard prove to be a useful genetic resource in the short term, or the seeds that enable mankind to get back on its feet after a catastrophe, food is certain to be a vital ingredient of humanity's future.

ACKNOWLEDGMENTS

As books about food go, this is an unusual one because it says very little about the taste of food or the joy of eating. Given my focus on the "nonfood" uses of food, the reader might easily conclude that I am only concerned with food's anthropological or geopolitical significance, and that I am not terribly interested in cooking or eating. Nothing could be further from the truth; and appropriately enough many of the people who helped me while I was writing this book did so over a meal. Toby Mundy of Atlantic Books crystallized my thoughts and proposed the title over lunch in Soho. George Gibson of Walker & Company embraced the idea over afternoon tea. I had constructive discussions with James Crabtree over sushi, Andreas Kluth over lunch at Zuni in San Francisco, Sarah Murray over coffee and cakes, and Paul Abrahams over lunch at the Garrick Club. Oliver Morton and Nancy Hynes helped me shape my ideas over several home-cooked meals.

Vital roles were played by Katinka Matson, my agent, who helped me devise the idea for this book, and Jackie Johnson, my editor, who fine-tuned the recipe. Expert advice was provided by Michael Pollan, Tim Harford, Adrian Williams, Matt Ridley, Felipe Fernández-Armesto, and Marion Nestle. I would also like to express my gratitude to the many other people who helped things along during the writing process, including Tamzin Booth, Edward McBride, John Parker, Ann Wroe, Edward Carr, and Geoffrey Carr at the *Economist*; Fitzroy Somerset; Endymion Wilkinson; Tom Moultrie and Kathryn Stinson; Tim Coulter and Maureen Stapleton (thank you for the corn and peanut soup); Zoe and Patrick Ayling; Anneliese St-Amour; Cristiana Marti (a magician with deep-fried zucchini flowers); Kate Farquhar; Nick Powell; Chester Jenkins; Stephan Somogyi; Lee McKee; and

Virginia Benz and Joe Anderer, with whom I have enjoyed many memorable meals over the years.

Last but certainly not least I would like to thank my children, Ella and Miles, and my wife, Kirstin, who was the first to encourage me to take on the topic of food—and to whom I hereby vow never to mention turnips or the Norfolk four-course rotation ever again.

NOTES

Part I

The account of the origins and domestication of maize follows Fussell, *The Story of Corn*; Warman, *Corn and Capitalism*; and Doebley, "The Genetics of Maize Evolution." The discussion of the domestication of rice and wheat, and of domestication more widely, follows Diamond, "Evolution, Consequences and Future of Plant and Animal Domestication"; Cowan and Watson, *The Origins of Agriculture*; and Needham and Bray, *Science and Civilisation in China*. For food-related creation myths, see Gray, *The Mythology of All Races*, and Visser, *Much Depends on Dinner*. The impact of farming on human health is discussed in Cohen, *Health and the Rise of Civilization*, and Manning, *Against the Grain*. The nature and impact of the spread of agriculture in Europe is discussed in Pinhasi, Fort, and Ammerman, "Tracing the Origin and Spread of Agriculture in Europe," and Dupanloup, Bertorelle, Chikhi, and Barbujani, "Estimating the Impact of Prehistoric Admixture on the Genome of Europeans."

Part II

The social structure of hunter-gatherer bands is discussed in Sahlins, *Stone Age Economics*, and Lee, *The !Kung San*. The transition from egalitarian hunter-gatherers to settled and socially stratified city-dwellers is discussed in Bellwood, *First Farmers*; Bender, "Gatherer-Hunter to Farmer: A Social Perspective"; Gilman, "The Development of Social Stratification in Bronze Age Europe"; Wenke, *Patterns in Prehistory*; Hayden, *Archaeology*; and Johnson and Earle, *The Evolution of Human Societies*. The account of Inca fertility rituals follows Bauer, "Legitimization of the State in Inca Myth and Ritual." A masterful comparative account of the emergence and structure of the earliest civilizations is provided by Trigger, *Understanding Early Civilizations*.

Part III

For spice-related myths, see Dalby, *Dangerous Tastes*. The origins and history of the spice trade are discussed by Dalby, *Food in the Ancient World from A to Z*; Schivelbusch, *Tastes of Paradise*; Keay, *The Spice Route*; Turner, *Spice*; and Miller, *The Spice Trade of the Roman Empire*. For the relationship between spices and trade, see Curtin, *Cross-Cultural Trade in World History*. For the roles of spices in spreading and supposedly warding off the Black Death, see Ziegler, *The Black Death*; Deaux, *The Black Death, 1347*; and Herlihy, *The Black Death and the Transformation of the West*. The fall of Constantinople is discussed in Crowley, *Constantinople*. Voyages of Columbus and Vasco da Gama are described in Fernández-Armesto, *Columbus*; Subrahmanyam, *The Career and Legend of Vasco da Gama*; Keay, *The Spice Route*; Turner, *Spice*; and Boorstin, *The Discoverers*. The impact of Vasco da Gama (and Zheng He) on European spice prices is discussed in O'Rourke and Williamson, "Did Vasco da Gama Matter for European Markets?" The structure of Indian Ocean trade is described in Chaudhuri, *Trade and Civilisation in the Indian Ocean*. The origins of European empires are discussed in Scammell, *The World Encompassed*. The local-food debate is examined in Murray, *Moveable Feasts*, and by innumerable bloggers online.

Part IV

The story of King Charles's pineapple follows Beauman, *The Pineapple*. European nations' competition in economic botany, and the origins of botanical gardens, are discussed in Brockway, *Science and Colonial Expansion*, and Drayton, *Nature's Government*. The transfer of maize and potatoes to the Old World are discussed in Ho, "The Introduction of American Food Plants into China"; Langer, "Europe's Initial Population Explosion"; and Langer, "American Foods and Europe's Population Growth 1750–1850." The account of transfer of sugar to the New World, and the proto-industrial nature of sugar production, follows Landes, *The Wealth and Poverty of Nations*; Mintz, *Sweetness and Power*; Hobhouse, *Seeds of Change*; Daniels and Daniels, "The Origin of the Sugarcane Roller Mill"; Higman, "The Sugar Revolution"; and Fogel, *Without Consent or Contract*. The history and impact of the potato are discussed in Salaman, *The History and Social Influence of the Potato*; Reader, *Propitious Esculent*; and McNeill, "How the Potato Changed the World's History." The discussion of the role of new foodstuffs and agricultural techniques in triggering the Industrial Revolution draws upon Malanima, "Energy Crisis and Growth 1650–1850"; Thomas, *The Industrial Revolution and the Atlantic Economy*; Pomeranz, *The Great Divergence*; Thomas, "Escaping from Constraints: The Industrial Revolution in a Malthusian Context"; Steinberg, "An Ecological Perspective on the Origins of Industrialization"; Wrigley, *Poverty, Progress and Population*; Wrigley, *Continuity, Chance and Change*; Jones, "Agricultural Origins of Industry"; and Jones, "Environment, Agriculture, and Industrialization in Europe." The account of the potato famine follows Reader, *Propitious Esculent*, and Hobhouse, *Seeds of Change*.

Part V

Military logistics in the ancient world are discussed by Engels, *Alexander the Great and the Logistics of the Macedonian Army*; Roth, *The Logistics of the Roman Army at War*; Clausen, "The Scorched Earth Policy, Ancient and Modern"; and Erdkamp, *Hunger and the Sword*. The role of logistics in the Revolutionary War is discussed by Tokar, "Logistics and the British Defeat in the Revolutionary War," and Bowler, *Logistics and the Failure of the British Army in America*. For a broad overview of the evolution of military logistics, see van Creveld, *Supplying War*, and Lynn, *Feeding Mars*. The account of Napoleon's rise and fall follows Rothenberg, *The Art of Warfare in the Age of Napoleon*; Nafziger, *Napoleon's Invasion of Russia*; Asprey, *The Rise and Fall of Napoleon Bonaparte*; Schom, *Napoleon Bonaparte*; and Riehn, *1812: Napoleon's Russian Campaign*. The role of logistics in the Civil War is discussed in Moore, "Mobility and Strategy in the Civil War." The account of the development of canned food follows Shephard, *Pickled, Potted and Canned*. The account of the Soviet famine of 1932–33 follows Ellman, "The

Role of Leadership Perceptions and of Intent in the Soviet Famine of 1931–1934"; Ellman, "Stalin and the Soviet Famine of 1932–33 Revisited"; and Dalrymple, "The Soviet Famine of 1932–1934." The great Chinese famine is discussed in Smil, "China's Great Famine: 40 Years Later," and Becker, *Hungry Ghosts*. The role of food shortages in the collapse of the Soviet Union is described in Gaidar, *Collapse of an Empire*. For an account of the sugar boycott see Wroe, "Sick with Excess of Sweetness."

Part VI

The account of the development of the Haber-Bosch process follows Smil, *Enriching the Earth*; Erisman, Sutton, Galloway, Klimont, and Winiwarter, "How a Century of Ammonia Synthesis Changed the World"; and Smil, "Nitrogen and Food Production: Proteins for Human Diets." The green revolution and its impact are discussed in Evans, *Feeding the Ten Billion*; Easterbrook, "Forgotten Benefactor of Humanity"; Evenson and Gollin, "Assessing the Impact of the Green Revolution, 1960 to 2000"; Webb, "More Food, But Not Yet Enough"; and Stuertz, "Green Giant." The relationship between agricultural productivity and economic development is discussed in Gulati, Fan, and Dalafi, "The Dragon and the Elephant: Agricultural and Rural Reforms in China and India"; Timmer, "Agriculture and Pro-Poor Growth: An Asian Perspective"; Delgado, Hopkins, and Kelly, "Agricultural Growth Linkages in Sub-Saharan Africa"; Fan, Hazell, and Thorat, "Government Spending, Growth, and Poverty: An Analysis of Interlinkages in Rural India"; Gollin, Parente, and Rogerson, "The Food Problem and the Evolution of International Income Levels"; Gollin, Parente, and Rogerson, "The Role of Agriculture in Development"; and Doepke, "Growth Takeoffs." Demographic transition is discussed in Doepke, "Accounting for Fertility Decline During the Transition to Growth." The relationship between nitrogen inputs and yields, and the scope for a switch to less chemical-intensive farming, is discussed in Smil, *Enriching the Earth*.

SOURCES

Asprey, Robert B. *The Rise and Fall of Napoleon Bonaparte*. London: Little, Brown, 2001.

Bauer, Brian S. "Legitimization of the State in Inca Myth and Ritual." *American Anthropologist* 98, no. 2 (June 1996): 327–37.

Beauman, Fran. *The Pineapple: King of Fruits*. London: Chatto & Windus, 2005.

Becker, Jasper. *Hungry Ghosts: China's Secret Famine*. London: John Murray, 1996.

Bellwood, Peter S. *First Farmers: The Origins of Agricultural Societies*. Oxford: Blackwell, 2005.

Bender, Barbara. "Gatherer-Hunter to Farmer: A Social Perspective." *World Archaeology* 10, no. 2 (1978): 204–22.

Boorstin, Daniel J. *The Discoverers*. New York: Random House, 1983.

Bowler, Arthur. *Logistics and the Failure of the British Army in America, 1775–1783*. Princeton, New Jersey: Princeton University Press, 1975.

Brockway, Lucile H. *Science and Colonial Expansion: The Role of the British Royal Botanic Gardens*. New York: Academic Press, 1979.

Chaliand, Gérard. *The Art of War in World History*. Berkeley: University of California Press, 1994.

Chanda, Nayan. *Bound Together: How Traders, Preachers, Adventurers, and Warriors Shaped Globalization*. New Haven: Yale University Press, 2007.

Chaudhuri, Kirti. *Trade and Civilisation in the Indian Ocean: An Economic History from the Rise of Islam to 1750*. Cambridge: Cambridge University Press, 1985.

Clausen, Wendell. "The Scorched Earth Policy, Ancient and Modern." *The Classical Journal* 40, no. 5 (February 1945): 298–99.

Clausewitz, Carl von. *On War*. London: Trübner, 1873.

Cohen, Mark. *Health and the Rise of Civilization*. New Haven: Yale University Press, 1989.

Cowan, C. Wesley, and Patty Watson, eds. *The Origins of Agriculture: An International Perspective*. Washington, D.C.: Smithsonian Institution Press, 1992.

Crowley, Roger. *Constantinople: The Last Great Siege, 1453*. London: Faber, 2005.

Curtin, Philip. *Cross-Cultural Trade in World History*. Cambridge: Cambridge University Press, 1984.

D'Souza, Frances. "Democracy as a Cure for Famine." *Journal of Peace Research* 31, no. 4 (November 1994): 369–73.

Dalby, Andrew. *Dangerous Tastes: The Story of Spices*. London: British Museum Press, 2000.

———. *Food in the Ancient World from A to Z*. London: Routledge, 2003.

———. *Siren Feasts: A History of Food and Gastronomy in Greece*. London: Routledge, 1996.

Dalrymple, Dana. "The Soviet Famine of 1932–1934." *Soviet Studies* 15, no. 3 (January 1964): 250–84.

Daniels, John, and Christian Daniels. "The Origin of the Sugarcane Roller Mill." *Technology and Culture* 29, no. 3 (July 1988): 493–535.

Davis, Johnny. "Svalbard Global Seed Vault: Ark of the Arctic." *Daily Telegraph*, February 16, 2008.

Deaux, George. *The Black Death, 1347*. London: Hamilton, 1969.

Delgado, L. C., J. Hopkins, and V. A. Kelly. "Agricultural Growth Linkages in Sub-Saharan Africa." International Food Policy Research Institute Research Report No. 107. Washington, D.C.: International Food Policy Research Institute, 1998.

Diamond, Jared. "Evolution, Consequences and Future of Plant and Animal Domestication." *Nature* 418 (August 8, 2002): 700–707.

———. *Guns, Germs, and Steel: The Fates of Human Societies*. New York: W.W. Norton, 1997.

———. "The Worst Mistake in the History of the Human Race." *Discover*, May 1987: 64–66.

Doebley, John. "The Genetics of Maize Evolution." *Annual Review of Genetics* 38 (2004): 37–59.

Doepke, Matthias. "Accounting for Fertility Decline During the Transition to Growth." *Journal of Economic Growth* 9, no. 3 (2004): 347–83.

————. "Growth Takeoffs." UCLA Department of Economics Working Paper 847 (2006).

Drayton, Richard. *Nature's Government: Science, Imperial Britain, and the "Improvement" of the World*. New Haven: Yale University Press, 2000.

Dupanloup, Isabelle, Giorgio Bertorelle, Lounès Chikhi, and Guido Barbujani. "Estimating the Impact of Prehistoric Admixture on the Genome of Europeans." *Molecular Biology and Evolution* 21, no. 7 (2004): 1361–72.

Easterbrook, Gregg. "Forgotten Benefactor of Humanity." *The Atlantic* 279, no. 1 (January 1997): 75–82.

Ellman, Michael. "The Role of Leadership Perceptions and of Intent in the Soviet Famine of 1931–1934." *Europe-Asia Studies* 57, no. 6 (September 2005): 823–41.

————. "Stalin and the Soviet Famine of 1932–33 Revisited." *Europe-Asia Studies* 59, no. 4 (June 2007): 663–93.

Engels, Donald W. *Alexander the Great and the Logistics of the Macedonian Army*. Berkeley, Los Angeles, and London: University of California Press, 1978.

Erdkamp, Paul. *Hunger and the Sword: Warfare and Food Supply in Roman Republican Wars (264–30 BC)*. Amsterdam: Gieben, 1998.

Erisman, J. W., M. A. Sutton, J. Galloway, Z. Klimont, and W. Winiwarter. "How a Century of Ammonia Synthesis Changed the World." *Nature Geoscience* 1 (2008): 636–39.

Evans, Lloyd T. *Feeding the Ten Billion: Plants and Population Growth*. Cambridge: Cambridge University Press, 1998.

Evenson, R. E., and D. Gollin. "Assessing the Impact of the Green Revolution, 1960 to 2000." *Science* 300 (May 2, 2003): 758–62.

Fan, S., P. Hazell, and S. Thorat "Government Spending, Growth, and Poverty: An Analysis of Interlinkages in Rural India." EPTD Discussion Paper No. 33. Washington, D.C.: International Food Policy Research Institute, 1998.

Fernández-Armesto, Felipe. *Columbus*. Oxford: Oxford University Press, 1991.

————. *Food: A History*. London: Macmillan, 2001.

Fogel, Robert William. *Without Consent or Contract: The Rise and Fall of American Slavery*. New York: W.W. Norton, 1989.

Foster, C., K. Green, M. Bleda, P. Dewick, B. Evans, A. Flynn, and J. Mylan. "Environmental Impacts of Food Production and Consumption: A Report to the Department for Environment, Food and Rural Affairs." London: DEFRA, 2006.

Fussell, Betty. *The Story of Corn*. New York: Knopf, 1992.

Gaidar, Yegor. *Collapse of an Empire: Lessons for Modern Russia*. Washington, D.C.: Brookings Institution Press, 2007.

Garnsey, P. *Food and Society in Classical Antiquity*. Cambridge: Cambridge University Press, 1999.

Gilman, Antonio. "The Development of Social Stratification in Bronze Age Europe." *Current Anthropology* 22, no. 1 (1981): 1–23.

Gollin, Douglas, Stephen L. Parente, and Richard Rogerson. "The Food Problem and the Evolution of International Income Levels." Working Papers 899, Economic Growth Center, Yale University, 2004.

———. "The Role of Agriculture in Development." *American Economic Review* 92, no. 2 (2002): 160–64.

Gray, Louis Herbert, ed. *The Mythology of All Races*. New York: Cooper Square Press, 1978.

Gulati, Ashok, Shenggen Fan, and Sara Dalafi. "The Dragon and the Elephant: Agricultural and Rural Reforms in China and India." International Food Policy Research Institute, MTID discussion paper 87. Washington, D.C.: International Food Policy Research Institute, 2005.

Gunn, Geoffrey. *First Globalization: The Eurasian Exchange, 1500–1800*. Lanham, Maryland: Rowman & Littlefield Publishers, 2003.

Hampl, Jeffrey, and William Hampl. "Pellagra and the Origin of a Myth: Evidence from European Literature and Folklore." *Journal of the Royal Society of Medicine*, 90, no. 11 (1997): 636–39.

Hayden, Brian D. *Archaeology: The Science of Once and Future Things*. New York: W.H. Freeman, 1993.

Herlihy, David. *The Black Death and the Transformation of the West*. Cambridge, Massachusetts: Harvard University Press, 1997.

Higman, B. W. "The Sugar Revolution." *The Economic History Review*, new series, 53, no. 2 (May 2000): 213–36.

Ho, Ping-Ti. "The Introduction of American Food Plants into China." *American Anthropologist*, new series 57, no. 2, part 1 (April 1955): 191–201.

Hobhouse, Henry. *Seeds of Change: Five Plants that Transformed Mankind.* London: Sidgwick & Jackson, 1985.

Johnson, Allen W., and Timothy Earle. *The Evolution of Human Societies: From Foraging Group to Agrarian State.* Stanford, California: Stanford University Press, 2000.

Jones, E. L. "Agricultural Origins of Industry." *Past and Present Society* 40, no. 1 (July 1968): 58–71.

———. "Environment, Agriculture, and Industrialization in Europe." *Agricultural History* 51, no. 3 (July 1977): 491–502.

Keay, John. *The Spice Route: A History.* London: John Murray, 2005.

Keegan, John. *A History of Warfare.* London: Hutchinson, 1993.

Kiple, Kenneth, and Kriemhild Ornelas, eds. *Cambridge World History of Food.* Cambridge: Cambridge University Press, 2000.

Kiple, Kenneth F. *A Movable Feast: Ten Millennia of Food Globalisation.* Cambridge: Cambridge University Press, 2007.

Landes, David. *The Wealth and Poverty of Nations: Why Some are So Rich and Some So Poor.* London: Little, Brown, 1998.

Langer, William. "American Foods and Europe's Population Growth 1750–1850." *Journal of Social History* 8, no. 2 (Winter 1975): 51–66.

———. "Europe's Initial Population Explosion." *American Historical Review* 69, no. 1 (October 1963): 1–17.

Lee, Richard Borshay. *The !Kung San: Men, Women and Work in a Foraging Society.* Cambridge: Cambridge University Press, 1979.

Lehane, Brendan. *The Power of Plants.* London: John Murray, 1977.

Lehner, Ernst, and Johanna Lehner. *Folklore and Odysseys of Food and Medicinal Plants.* London: Harrap, 1973.

Lynn, John A., ed. *Feeding Mars: Logistics in Western Warfare from the Middle Ages to the Present.* Boulder, Colorado: Westview Press, 1993.

Malanima, Paolo. "Energy Crisis and Growth 1650–1850: The European Deviation in a Comparative Perspective." *Journal of Global History* 1, no. 1 (2006): 101–21.

Malthus, Thomas. *An Essay on the Principle of Population.* London: J. Johnson, 1803.

Manning, Richard. *Against the Grain: How Agriculture Has Hijacked Civilization.* New York: North Point Press, 2004.

Marks, Robert. *The Origins of the Modern World: A Global and Ecological Narrative from the Fifteenth to the Twenty-first Century.* 2nd edition. Lanham, Maryland: Rowman & Littlefield, 2006.

McGee, Harold. *McGee on Food & Cooking: An Encyclopedia of Kitchen Science, History and Culture.* London: Hodder and Stoughton, 2004.

McNeill, William H. "How the Potato Changed the World's History." *Social Research* 66, no. 1 (Spring 1999): 67–83.

Michalowski, Piotr. "An Early Dynastic Tablet of ED Lu A from Tell Brak (Nagar)." *Cuneiform Digital Library Journal* 2003:3, http:cdli.ucla.edu/pubs/cdlj/2003/cdlj2003_003.html.

Miller, J. Innes. *The Spice Trade of the Roman Empire, 29 BC to AD 641.* Oxford: Clarendon Press, 1969.

Mintz, Sidney Wilfred. *Sweetness and Power: The Place of Sugar in Modern History.* New York: Viking, 1985.

Mithen, Steven. *After the Ice: A Global Human History 20,000–5,000 BC.* Cambridge, Massachusetts: Harvard University Press, 2004.

Moore, John. "Mobility and Strategy in the Civil War." *Military Affairs* 24, no. 2, Civil War Issue (Summer 1960): 68–77.

Murray, Sarah. *Moveable Feasts: From Ancient Rome to the 21st Century, the Incredible Journeys of the Food We Eat.* London: Macmillan, 2007.

Nafziger, George. *Napoleon's Invasion of Russia.* Novato, California: Presidio, 1998.

Needham, Joseph, and Francesca Bray. *Science and Civilisation in China* vol. 6, *Biology and Biological Technology,* part 2, *Agriculture.* Cambridge: Cambridge University Press, 1984.

Newman, Lucile F., ed. *Hunger in History: Food Shortage, Poverty, and Deprivation.* Oxford: Blackwell, 1990.

O'Rourke, Kevin, and Jeffrey Williamson. "Did Vasco da Gama Matter for European Markets?" National Bureau of Economic Research, Working Paper 11884, 2005.

Ogilvie, Brian. "The Many Books of Nature: Renaissance Naturalists and Information Overload." *Journal of the History of Ideas* 64, no. 1 (January 2003): 29–40.

Pinhasi, Ron, Joaquim Fort, and Albert Ammerman. "Tracing the Origin and Spread of Agriculture in Europe." *PLoS Biology* 3, no. 12 (2005): e410.

Pollan, Michael. *The Omnivore's Dilemma: A Natural History of Four Meals.* New York: Penguin Press, 2006.

Pomeranz, Kenneth. *The Great Divergence: China, Europe, and the Making of the Modern World Economy.* Princeton, New Jersey: Princeton University Press, 2000.

Reader, John. *Propitious Esculent: The Potato in World History.* London: William Heinemann, 2008.

Riehn, Richard K. *1812: Napoleon's Russian Campaign.* New York: John Wiley & Sons, 1991.

Rotberg, Robert I., and Theodore K. Rabb, eds. *Hunger and History: The Impact of Changing Food Production and Consumption Patterns on Society.* Cambridge: Cambridge University Press, 1985.

Roth, Jonathan. *The Logistics of the Roman Army at War (264 BC–AD 235).* Leiden, Netherlands: Brill, 1998.

Rothenberg, Gunther Erich. *The Art of Warfare in the Age of Napoleon.* Bloomington: Indiana University Press, 1977.

Sahlins, Marshall David. *Stone Age Economics.* London: Tavistock Publications, 1974.

Salaman, Redcliffe N. *The History and Social Influence of the Potato.* Cambridge: Cambridge University Press, 1949.

Scammell, G. V. *The World Encompassed: The First European Maritime Empires, c.800–1650.* Berkeley and Los Angeles: University of California Press, 1982.

Schivelbusch, Wolfgang. *Tastes of Paradise: A Social History of Spices, Stimulants, and Intoxicants.* Translated by David Jacobson. New York: Vintage, 1992.

Schom, Alan. *Napoleon Bonaparte.* New York: HarperCollins, 1997.

Sen, Amartya. "Democracy as a Universal Value." *Journal of Democracy* 10, no. 3 (1999): 3–17.

Shephard, Sue. *Pickled, Potted and Canned: The Story of Food Preserving.* London: Headline, 2000.

Smil, Vaclav. "China's Great Famine: 40 Years Later." *British Medical Journal* 319, no. 7225 (1999): 1619–21.

———. *Enriching the Earth: Fritz Haber, Carl Bosch, and the Transformation of World Food Production.* Cambridge, Massachusetts: MIT Press, 2004.

————. "Nitrogen and Food Production: Proteins for Human Diets." *Ambio* 31, no. 2 (March 2002): 126–31.

Smith, Adam. *An Inquiry into the Nature and Causes of the Wealth of Nations*. Hartford: Oliver D. Cooke, 1804.

Spaull, C. "The Hekanakhte Papers and Other Early Middle Kingdom Documents, by T. G. H. James." (Review) *Journal of Egyptian Archaeology* 49 (1963): 184–86.

Steinberg, Theodore. "An Ecological Perspective on the Origins of Industrialization." *Environmental Review* 10, no. 4 (Winter 1986): 261–76.

Stuertz, Mark. "Green Giant." *Dallas Observer*, December 5, 2002.

Subrahmanyam, Sanjay. *The Career and Legend of Vasco da Gama*. Cambridge: Cambridge University Press, 1997.

Svalbard Global Seed Vault official Web site: http://www.regjeringen.no/en/dep/lmd/campain/svalbard-global-seed-vault.html.

The State of Food and Agriculture 2003–2004. Rome: Food and Agriculture Organization of the United Nations, 2004.

Thomas, Brinley. "Escaping from Constraints: The Industrial Revolution in a Malthusian Context." *Journal of Interdisciplinary History* 15, no. 4 (Spring 1985): 729–53.

————. *The Industrial Revolution and the Atlantic Economy: Selected Essays*. London: Routledge, 1993.

Timmer, Peter. "Agriculture and Pro-Poor Growth: An Asian Perspective." Center for Global Development, Working Paper 63, July 2005.

Tokar, John. "Logistics and the British Defeat in the Revolutionary War." *Army Logistician* 31, no. 5 (September–October 1999): 42–47.

Toussaint-Samat, Maguelonne. *A History of Food*. Oxford: Blackwell, 1992.

Trigger, Bruce G. *Understanding Early Civilizations*. Cambridge: Cambridge University Press, 2003.

Turner, Jack. *Spice: The History of a Temptation*. New York: Knopf, 2004.

Van Creveld, Martin. *Supplying War: Logistics from Wallenstein to Patton*. Cambridge: Cambridge University Press, 1977.

Visser, Margaret. *Much Depends on Dinner: The Extraordinary History and Mythology, Allure and Obsessions, Perils and Taboos of an Ordinary Meal*. New York: Grove Press, 1987.

Warman, Arturo. *Corn and Capitalism: How a Botanical Bastard Grew to Global Dominance.* Translated by Nancy L. Westrate. Chapel Hill: University of North Carolina Press, 2003.

Webb, Patrick. "More Food, But Not Yet Enough: 20th Century Successes in Agriculture Growth and 21st Century Challenges." Friedman School of Nutrition, Tufts University, Food Policy and Applied Nutrition Program Discussion Paper 38, 2008.

Wenke, Robert J. *Patterns in Prehistory: Humankind's First Three Million Years.* New York: Oxford University Press, 1990.

Wittfogel, Karl August. *Oriental Despotism: A Comparative Study of Total Power.* New Haven: Yale University Press, 1959.

Wrigley, Edward Anthony. *Continuity, Chance and Change: The Character of the Industrial Revolution in England.* Cambridge: Cambridge University Press, 1988.

——. *Poverty, Progress and Population.* Cambridge: Cambridge University Press, 2004.

Wroe, Anne. "Sick with Excess of Sweetness." *Economist,* December 19, 2006.

Ziegler, Philip. *The Black Death.* London: Collins, 1969.

INDEX

Abu Hureyra (Syria), 20

Africa. *See also specific countries*
  circumnavigation of, 69, 77,
    91–93
  green revolution in, 234–35, 236
  hunter-gatherers, 16–17, 33–36
  slaves from, 114–15
  spread of Islam, 77

agricultural productivity. *See also*
  green revolution
  application of nitrogen and, 202–6
  cereal crop mutation and
    selection, 9–11
  dwarf crop varieties and, 213–20
  and the emergence of strong
    leaders, 41–44
  and fall of the Soviet Union,
    188–92
  and industrialization, 130–31,
    177–79, 182–83, 221–22
  maize mutation and selection, 5–9
  Malthusian trap, 124–28, 129–30,
    139–40, 217–18, 226–29
  Stalin's collective farms and,
    177–82

agriculture
  conservation tillage, 236–37
  creation myths and, 13–15
  disadvantages of switch from
    hunter-gatherer lifestyle to
    farming, 16–19

environmental impact of, 27
extent of wheat, rice, and maize
  farming, 12(map)
and inequality, 55–56
interdependence of crops and
  humans, 7, 25–27
loss of association with the land,
  57–59
organic farming, 232–33, 236, 237
origins of, 4, 19–22
spread of farming and farmers,
  22–25
as technology, 3–4, 26–27

Ailly, Pierre d', 86

al-Biruni. *See* Biruni, al-

Alexander the Great, 146–48

Alexandria Tariff, 65

allspice, 90

American Civil War, 162, 163–68

American Revolutionary War, 149–51

ammonia synthesis, 199–200, 206–12

Anatolians, 24

animal domestication, 11, 25

Appert, Nicolas, 160–61, 162, 163

Arab traders, 63–65, 72, 77–78.
  See also Islam

The Art of Preserving All Kinds of
  Animal and Vegetable
  Substances for Several Years
  (Appert), 161

Austerlitz, battle of, 154–55

A Note on the Author

TOM STANDAGE is business editor at the *Economist* and the author of *A History of the World in 6 Glasses* (a *New York Times* bestseller), *The Turk*, *The Neptune File*, and *The Victorian Internet*, described by the *Wall Street Journal* as a "dot-com cult classic." *The Victorian Internet* was made into a documentary film, *How the Victorians Wired the World*. He has written about science and technology for numerous magazines and newspapers, including *Wired*, the *Guardian*, the *Daily Telegraph*, and the *New York Times*. Standage holds a degree in engineering and computer science from Oxford University, and he is the least musical member of a musical family. He lives in London, England, with his wife, daughter, and son.